A Case Study on Indian Agriculture

A Case Study on Indian Agriculture

Manish Dubey

CYBER TECH PUBLICATIONS
4264/3, Ansari Road, Daryaganj, New Delhi-110002 (India)
Ph.: 011-23244078, 011-43559448 Fax: 011-23280028
E-Mail: cyberpublicationsdelhi@yahoo.com
Website: www.cybertechpublications.com

A CASE STUDY ON INDIAN AGRICULTURE

Manish Dubey

First Edition 2010

Published by : Asian Offset Printers

G.S. Rawat for **Cyber Tech Publications**
4264/3, Ansari Road, Daryaganj, New Delhi-110002 (India)
Ph.: 011-23244078, 011-43559448 Fax: 011-23280028
E-Mail: cyberpublicationsdelhi@yahoo.com
Website: www.cybertechpublications.com

Preface

The emerging and increasing market opportunities for organics are conducive for the adoption of organic agriculture among small scale farmers in India. However, some studies suggest that major constraints exist for small scale farmers to reach these markets and at the same time secure a price premium, while other evidence suggests that it is possible to support farmers to access organic markets and benefit, especially in terms of increased premiums. For the International Fund for Agricultural Development (IFAD) to consider including organic farming in the projects it supports in these countries, it is essential to understand the factors that allow small farmers to resolve the most common problems related to the production and marketing of organic agricultural products.

The farm mechanization has been well received world over as one of the important elements of modernization of agriculture. In India, though there has been a considerable progress of mechanization in agriculture, its spread has been in the most uneven manner. Hitherto, governmental efforts have mostly been confined to the promotion of manual and animal-drawn tools and implements.

Thus, considering the expanding needs of the agricultural mechanization in the country, it has become necessary to ensure that the requirements of the farmers are duly met in terms of the local needs, quantity and quality of inputs and social benefits. This can only be accomplished by taking definite policy measures and the strategic planning. Keeping abreast with the recent advances & development in agricultural mechanization taking place all around, upgradation of technology is needed to be a continuous process.

— *Author*

CONTENTS

1

INTRODUCTION

This chapter focuses on the regulatory structures that have provided the context for the debate surrounding the introduction of gene technology into Indian agriculture and shows how protests against its introduction have influenced and been influenced by the evolution of international negotiations, national regulatory procedures and local interests. In the 15 years since biotechnology became a priority area for Indian policy makers there have been a series of crucial events that have deeply influenced thinking on biotechnology and the role of genetic engineering (GE) in agricultural production. The most important of these have been the multilateral negotiations that accompanied the agreement on Trade Related Aspects of Intellectual Property Rights (TRIPs), the formalization of the Cartagena Protocol on Biosafety (CPB), the threatened introduction of 'terminator seed' and the arrival of Bacillus thuringiensis (Bt) cotton.

CRITICISMS AND DOUBTS

Biotechnology has been received with a considerable degree of scepticism in Indian agricultural circles. Some critics questioned whether it could make any significant contribution to increasing food security, while others warned that it might actually threaten food supplies by stimulating the production of cash crops. There was also the possibility that it might adversely affect agricultural exports by encouraging the developments of synthetic substitutes for commodities such as sugar, an important Indian export.

Biotechnology was also associated with the private sector becoming increasingly involved in the provision of seeds and other planting material. Government policy in India is committed to ensuring food security. Public sector research focuses on both food and commercial crops and takes a balanced view of agriculture as a whole. The private sector, however, is primarily concerned with commercial crops and has largely neglected the orphan crops that form the mainstay of the diets of small and marginal farmers and agricultural labourers. The growing power of the private sector is seen by some to be undermining food security objectives.

GENETICALLY MODIFIED CROPS

It was the issues associated with terminator technology that provided a focus for concern about India's agro-biotechnology policy. Those involved in the ensuing debate came from all sectors of Indian society and included policy makers, providers of technology from publicly funded and private sector organisations, civil society groups and farmers organisations. In 1998, Delta and Pine Land Company (USA) and the United States Department of Agriculture (USDA) obtained a patent for a new genetic technology designed to produce sterile seeds. These seeds, soon dubbed 'terminator seeds', were intended to prevent farmers from reusing seeds obtained at harvest and were expected to increase the value of the proprietary seed owned by USA seed companies and open up new overseas markets.

India's connection with terminator technology came through the life science company Monsanto (USA) who after acquiring Delta and Pine Land declared its intention to market genetically modified (GM) seeds despite opposition from the Green lobby. The company's decision to acquire terminator technology provided the catalyst for the launch of a determined campaign against GM seeds by a coalition of civil society organizations and farmers' groups in India. Many of these groups were already lobbying against the decision to introduce Intellectual Property Rights (IPRs) into agriculture. In the multilateral negotiations that accompanied TRIPs and the Cartagena Protocol on Biosafety, the Indian Government had not only played an active role, it had also taken positions at critical moments in each set of negotiations. These positions were strongly influenced by the campaigns launched by those civil society organisations that sought to highlight the interests of Indian farmers and the need to protect the environment.

The key issue in meeting the commitments that the Indian Government had made under TRIPs to the World Trade Organization (WTO) was the type of IPRs that should be introduced. One of the options open to WTO members was the adoption of the framework provided by the International Union for the Protection of New Varieties of Plants (UPOV). The UPOV Convention, formalized in 1961, provided a system for granting Plant Breeders' Rights. In the past, the UPOV Convention had allowed exclusive rights to be granted to plant breeders for the varieties they had developed, and at the same time it also allowed traditional farmers to reuse seeds obtained from their annual harvest and to exchange these seeds with neighbours. The most recent amendment to UPOV, carried out in 1991 (UPOV'91), however, took away the freedom granted to farmers to exchange seeds with their farm neighbours or re-use seed for commercial purposes. UPOV'91, therefore, took away what farmers saw as one of their basic rights.

Small farmers in India had been vocal in their protest against UPOV'91 and the Indian government did not adopt the UPOV option. The discussion did, however, contribute to heightening farmer suspicion. It was against this background that the idea of terminator seed entered India. It was, therefore, no surprise that the farming community forthrightly rejected it. The government's firm position against the introduction of terminator seed was clearly influenced by civil society campaigns that were determined to prevent a strengthening of intellectual property laws that would make it easier for seed giants to dominate the local seed market.

GENETICALLY MODIFIED SEEDS

The controversy over the threat from terminator seeds had considerable fallout, jeopardizing field trials of a GM variety of cotton that Monsanto had been conducting since mid-1998. This variety of cotton had been developed by implanting a gene extracted from the naturally occurring Bt soil bacterium into the high yielding varieties of cotton already being cultivated in India. Monsanto claimed that this new variety of cotton would improve local cotton varieties and provide them with the potential to kill insect pests. This would result in substantial yield increases and, equally importantly, would offer what was presented as a more environmentally safe way of cultivating a crop that currently accounted for 40 per cent of the total volume of pesticide used in Indian agriculture each year.

In 1998, the Indian government had given the go-ahead for limited field trials of Bt cotton. Its view was that Monsanto's cotton variety could provide the necessary resistance to the bollworm pest and help prevent annual losses estimated to be worth up to US$ 190 million dollars. The Indian Council of Agricultural Research (ICAR) justified support for the adoption of Bt cotton on the grounds that the conventional crop required heavy applications of chemical pesticide in order to flourish. ICAR argued that conversion to Bt cotton would not only reduce environmental pollution, it would also reduce the burden of debt farmers were often forced to incur in order to purchase the pesticide needed to treat their cotton crop. However, one of the main points raised by those campaigning against the decision to introduce Bt cotton into India was that the process for approving field trials had not been carried out with the requisite level of transparency about the environmental risks of using GM seeds.

REGULATIONS

The regulation of genetically modified organisms (GMOs) in India has been subjected to the rules framed by the Ministry of Environment and Forests (MOEF) in 1989. These rules, which were part of the Environmental (Protection) Act of 1986, defined implementing structures for conducting research and for the commercial applications of GMOs. This regulatory structure consists of six committees: three are under the Department of Biotechnology (DBT), one under the MOEF and two operate at subfederal levels closer to the actual site of GM crop field trials. However, although the Government had established this comprehensive structure for regulating GMOs as early as 1989, long before the global community reflected on the need to pursue biosafety regulations, critics in India have raised questions about the way the system operates.

One of the more pertinent questions relates to the fact that this structure of six committees is largely redundant since only two of the committees seem to have any role to play. These are the Review Committee on Genetic Manipulation (RCGM) of the DBT and the Genetic Engineering Approval Committee (GEAC) of the MOEF. The functions of these sub-federal committees remain largely undefined, and the way in which rules have been framed for the committees concerned with the regulatory administration of GMOs have given cause for concern.

The issue of Bt field trials brought these concerns to the fore. The regulatory procedure followed by the RCGM in the case of Bt cotton was questioned on several grounds by civil society organizations.

First, field trials were approved by the RCGM after the farmers chosen for this purpose had sown the Monsanto seeds in their fields.

Second, there was criticism of the role played by the sub-federal committees in granting approvals for field trials at 40 locations in nine states.

Finally, the functioning of RCGM was called into question because changes in the rules of the committee were made when Bt trials were still in progress. This lack of transparency in granting clearance to the field trials of Bt cotton was the single most important factor behind the farmers' protests in the Southern states. Proponents of genetic engineering argued that civil society campaigns had spread misinformation and re-emphasised that the objective of promoting biotechnology was essentially to increase the viability of Indian farmers. Misinformation or not, the campaigns managed to put sufficient pressure on the government to prevent an early introduction of Bt cotton on the market.

GOVERNMENT DECISION-MAKING

The way in which Bt cotton was introduced into India illustrates the complex nature of the processes that are involved in governmental decision-making. In the past, vital decisions were often made and implemented before their implications were analysed. Today, we are witnessing a more active involvement of stakeholders before decisions are made on any particular issue. Many would insist that the lack of transparency still pervades but times are in fact changing.

In 2000, the three Southern Indian states of Karnataka, Andhra Pradesh (both scenes of farmer protest in 1998 and 1999) and Tamil Nadu gave explicit support for biotechnology and began to announce their biotechnology policies. This has provided the necessary mandate to the federal government to fulfil its unfinished task of completing the Bt cotton field trials. State level support for biotechnology has a deeper significance in the Indian context where the Constitution of India (which delegates powers between the central government and the states) defines agriculture as one of the issues that should be managed exclusively at state level.

It is becoming clear that the number of seed companies and State governments in favour of introducing GM seeds is increasing. The seed companies are anxious to introduce GM seeds as part of their corporate strategy, while state governments are more inclined to see these products in the context of their stated policy of modernizing agriculture. In July 2000, the Genetic Engineering Approval Committee of MOEF approved large-scale field trials of Bt cotton by the Monsanto affiliate Maharashtra Hybrid Seeds Co. Ltd. (Mahyco) for seed production and demonstration and to generate environmental safety data for crops under various agricultural and climatic conditions (see also the article in Monitor No. 44/45). Mahyco was to undertake open field trials on 85 hectares and seed production on 150 hectares. Most significantly the ICAR, which includes agricultural universities as well as public-funded research institutions, would be fully involved in monitoring seed production. The decision to involve the ICAR system was taken even as the Central Institute for Cotton Research, one of the publicly-funded agricultural research institutes, was unveiling plans for putting its own variant of Bt cotton on the market by the year 2002.

The domestic biotechnology industry was beginning to assert itself in India, after spending years in the shadow of corporates like Monsanto. A start was made in April 2001 with agriculture high on the agenda and crops such as cotton, rape and mustard under research.

However, potential users of GM seeds within the farming community as a whole still do not appear to be convinced of their potential. In December 2000, the Minister of Agriculture informed the Indian parliament that a final decision on the use of GM seeds had yet to be taken. The campaigns run by civil society organisations have certainly contributed to this decision but much more coordinated effort will be required if these concerns are to be translated into policy.

Yields and Incomes Increase with GM Crops

Indian farmer Vitthal Narayan Patil has been growing cotton for more than 20 years. Patil has heard discussions about the pros and cons of biotechnology, but since first using insect-protected (Bt) cotton in 2002, Patil has seen his family's income increase and livelihood improve.

"Before Bollgard, in the traditional seeds of cotton, our expenses were high and our yield was low. But with the introduction of Bollgard, we are saved from the havoc of the bollworms, which would affect 30 percent of the yield," says Patil, a large cotton and banana grower who plants more than half of his family's 200 acres (80 hectares) with biotechnology crops.

"A farmer would always want more yield. We hope for success. Whatever we do, we do it with an expectation of success. And this biotechnology has brought that success. There's no doubt about that," continues Patil.

The adoption of biotechnology in India has been rapid-indicating growers are realizing numerous pros and few cons with biotechnology. But Patil believes more should be done to educate growers and encourage biotechnology in India to keep farmers and farm families economically viable. "Today's times are very competitive. If we are getting more yield at low cost, then the farmers can survive in this competitive market....

"Here there is a farmer I've known for the last 15 years... who was struggling with his financial situation. I bought the Bollgard seeds for him, helped him to grow, and now there is a tremendous change in his financial situation. He has bought a tractor and his condition has improved," says Patil.

Higher yields and increased income from biotechnology in India help farm families improve their standard of living. Patil intends to use the increased income on education for his children. "Nowadays education is very expensive. If a child is to go to a good school, where he can get good education, there he'll need more money. A farmer cannot afford so much expense. Only if he gets good yield, gets good money, is when he can afford good education."

MODERN BIOTECHNOLOGY IN LDCS

Two popular assumptions justify the introduction of Genetically Modified (GM) seeds in agriculture. One is demographics, and especially the expectation that according to current rates of population growth in the poorest areas of the world by 2020 not enough food will be produced to feed everyone on the planet. The other is global warming as a result of traditional farming methods associated with the Green Revolution.

Therefore, part of the international media and civil society stresses that innovative and sustainable solutions are badly needed and high expectations surround the introduction of biotechnology in agriculture. It should be said that genetic modifications of plants is not an entirely new thing as it has been around already at least since the classification of species and the early experiments of Mendel. However, molecular assisted genetic analysis and research make it possible to synthetise in the laboratory seeds whose DNA has been modified to endow them with special traits. By transferring genes from one species to another it is possible to generate seeds by apoptosis, basically isolating certain desired characteristics, say to make them more resistant to environmental agents, while removing undesired traits.

However, the debate on the introduction of modern biotechnology in agriculture is highly polarised. Enthusiasts point to the various possibilities that GM seeds can bring about from increasing yields, enhancing quality, reducing environmental impact to producing crops with improved nutrition contents. While opponents are concerned about the long-term impact on human health and the ecosystem, the controversial findings of the empirical studies conduced to date and the risks related to the appropriation of foreign multinationals patenting traditional local crops and staple food. India, given its geography and demographics, is one of the countries that will be most affected by food deficiencies in the future and is stimulating the imagination of many for its potential to become a major user of such technology. Presently, despite nearly 70% of the population is involved in agriculture, dispersions within the public distribution system mean that even if enough food is actually produced, it does not always reach the most needy, at the time most needed.

After nearly three months or so of pilgrimage from the fields of Haryana, across the desert of Rajastan, then through the tea leave plantations of Kerala, the catholic state of Goa and the mountains of the Nilgiri Hills National Park in Tamil Nadu there have been many thoughts and reflections along the way.

The objective of this reportage is to isolate the crucial aspects of the debate about the introduction of GM seeds in India's agricultural as well as to provide a disenchanted view of the relevant issues surrounding the use of biotechnology in agribusiness and especially the role of Trade Protection of Intellectual Property Rights (TRIPS). In a nutshell, whereas TRIPS are an obligation to WTO member countries, their effect on knowledge, innovation and wealth in LDCs is still unclear. Our exploratory study of India's agricultural markets will start with an overview of the local reality, the identification of major players and actors involved and the international dimension framing these interactions. The macro-institutional framework complementary to the adoption and distribution of GM seeds will be analysed and improvements in this area of development policy will be proposed. Conclusions and suggestions follow.

THE REALITY OF INDIA

Cruising around the mustard fields of Rajastan, discovering the enchanted rice fields of Karnataka or finding your way through the jungle in the hilly north west part of Tamil Nadu is indeed a remarkable experience. India possesses a phenomenal variety of land types and climate zones with vast farming areas surrounded at times by high mountains or flat plains, which make it virtually perfect for the adoption of large scale farming as well as boutique organic products. As it is often heard while travelling around the Subcontinent: "Everything is possible in India!". Nevertheless, in many of the cases your correspondent came across, production was still relatively small scale, often despite concentration of land ownership, and peasants used rather antiquated and labour intensive means of production.

Maharshtra, Andra Pradesh, Kerala and Karantaka are the most active states in terms of both research and application of biotechnology in agriculture. Due to the presence of the high-tech hub of Bangalore, the state of Karnataka is in the most attractive position the become a leader in the field, advertising tax allowances and a low cost floor space for companies involved in genetic engineering, as well as pharmaceutical research and bio-informatics.

Karnataka hosts also the Original Rice Research Station in Mandia (about 100 kilometers away from Bangalore). A noticeable presence considering that rice is India's staple food, occupies about 43 million hectares of cultivated land and accounts for over 40% of food grain production of the entire country. Unfortunately it seems that to date Golden Rice (a variety of rice grain with enhanced beta-carotene-provitamin A-nutrient contents) is not yet widely available, despite its potential to be an effective nutrients transmitter for the population at large while at the same time is considered by experts the 'number one' cultivable crop in the world.

Likewise Karnataka is also the fourth largest cotton producing state in India, which foresees fruitful prospects for the introduction of Bollgard, an innovative cotton seed variety introduced by Monsanto, which is resistant to bollworms. Bollgard involves the insertion of a gene called Bacillus thuringiensis (Bt), a naturally occurring insecticidal toxin by a soil-borne bacterium, into the germoplasm of a native specie of cotton plant. Bollworms can cause losses as high as 90% of the harvest, and until the discovery of Bt the only way to fight it would be with pesticides and chemicals. Today, cotton growers use nearly half of the pesticides of the entire agricultural sector, thus Bt could provide a more natural counter measure. At the same time both public and private research centres are investigating other alternative methods, while the various Government Departments of Agriculture are in the hectic process of training farmers in the use of bio-fertilisers.

The current overarching trend to 'push' GM seeds use in India is probably a response to the sky-rising levels of chemical fertlisers and pesticide consumtpion of the 1970s and 1980s, during the period of agricultural modernisation known as the 'Green Revolution'. The Green Revolution can be summarised as follows: Indian scientists planted genetic materials within self-pollinated high-yield seeds variety, while farmers were trained to use chemical fertilisers and massive amounts of pesticides. It was a shock for your correspondent to learn during a visit to a secluded Toda village in the Natural Reserve of the Nilgiri Hills at some 2,500 meters of altitude above the sea level, that chemical agents and pesticides were a standard practice among farmers. According to a respondent: "Thanks to the Green Revolution food production increased and costs decreased, but today the presence of chemicals in the soil can be as much as 30 Kgs per hectare, this is high, but not as high as in Japan were it can reach up to 80 Kgs per hectare... when it is cheaper for the farmer to use chemicals instead of more innovative and less polluting methods, poison can be sweet too!". Unfortunately, the indiscriminate use of chemical fertilisers is decreasing soil life and humus, developing nutrient imbalances in plants, polluting the water and creating a number of other environmental hazards. Thus, improving the current situation requires a coordinated effort across the board to reduce the costs for alternative methods to take over.

India is changing, but it has been for many years, and in many respects it still is, a closed market. This means that a number of local companies operate in the agricultural sector, which includes also companies in the area of animal feed, pesticides and tractors and foreign presence is limited. According to data gathered by Pray and Basant (2002) the largest Indian firms by market share and R&D expenditure are Hindustan Lever, Rollis India, United Phosphorous and Bayer India. Tata ltd should also be included being a large conglomerate whose main business is in the automobiles sector, but owns also tea plantations and factories, consulting services as well as insurance companies and beach resorts. The foreign players are Monsanto, Du Pont and Aventis, which have been investing heavily in GM research since 1998 and have established export-oriented units in agreement with the government. Estimates might not be entirely reliable, but during the 1990s the private sector spent between $39 and $43 million on food and agricultural research (Pray & Basant, 2002:35).

Sadly, looking at the conditions of the laborers the general sense of backwardness inspires sorrow, and while richer farmers owned at least one tractor, many others still used a wooden plough carried by cows or camels. The first impression is that you would not expect land owners to be aware of the existence of genetically modified seeds. Suspicion that was soon to be disproved. During a conversation with a farmer, he would complain that after buying pest resistant seeds, he would not be able to plant again the seeds from the fruit that grew thereafter without developing a Frankenstein-like hybrid plant. Quite unusually, the plant stemmed all the varieties of the original crop simultaneously.

Thus, he had to buy new seeds again. Such an occurrence is known as transgenic double-cross hybrid, which means that since growers are given seeds from single-cross hybrids, as opposed to seeds of two inbred lines, these cannot be reproduced. Fear of copying from competitors and loss of sales is what worries seed suppliers, which then introduce genetic restrictions (or Genetic Use Restriction Technologies-GURTs) to reproducibility. Yet the consequences of such defensive methods for Indian (as well as other LDCs) farmers are of some importance and will be explored in more detail.

The International Dimension and its Interdependencies with the Local Context

Innovations and new plant varieties are of great importance to increase farmers' income and promote wealth creation in general, unfortunately purchasing new inputs or labor is often a drawback that constrain farmers' choices and preferences. Embedding GURTs are not the only steps that the innovators are considering to protect intellectual property rights in agriculture. The following documents provide some guidance:

- TRIPS and especially article 27.3(b),
- the UN Convention on Biological Diversity (CBD),
- the International Undertaking on Plant Genetic Resources [for food and agriculture] (IU) by the Food and Agriculture Organization of the United Nations,
- and the International Convention for the Protection of New Varieties of Plants (UPOV).

According to an Oxfam study on the effects of patenting life forms in agriculture (Tkatchenko, 2002:5) the coordination between these agreements remains uncertain. Tkatchenko (2002) points out that there is a discussion about a potential synergy between the IU and the CBD, but an agreement has not yet been reached. The position of the CBD itself is precarious since it has received little backing from international bodies. And non-signatory states, including the US, continue to assert the supremacy of the TRIPS over the CBD. Currently, membership to the World Trade Organisation (WTO) requires the institutionalisation of TRIPs; under the assumption that innovations developed in the North can un-problematically diffuse and prosper in the South as well. Beside the varying degrees of truthfulness of the previous assumption in relation to the application of knowledge intensive technologies such as GM seeds, consider the example of Monsanto patenting Chapati, a traditional Indian bread. What is going to happen to enforce such a patent remains still unknown, but it will certainly contribute little to increase the available income of farmers or to promote acceptance by the general public of patents around GM technology. Unpredictably, this has happened despite India has adopted a sui generis system that gives the country more autonomy in setting patent legislation, which will be discussed in the following section.

Hence, it is not difficult to understand the reason why a general feeling of mistrust emerges from the various respondents about the use of transgenic seeds and many have more or less openly declared their opposition. Yet the macro-institutional framework is moving, albeit slowly, for their legitimisation. Remember that the Indian seed sector hosts a number of small firms, but only few with large operations; Monsanto and Du Pont being the largest.

A possible risk coming from the present industry structure is that due to its fragmentation consensus over legislation would require a long and tedious process of negotiation among the various producers, while bigger players could have an easier life and greater impact and say on future decisions. For instance, while local newspapers acclaim that Second Green Revolution is taking place in seed production, some of Punjab's large farms contain state-of-the-art Research and Development laboratories that produce only for exports. On the other hand, GM seeds remain still unaffordable for the majority of Indian farmers, require the use of expensive bio-fertilisers and risk to perpetuate the spiral of indebtedness that affects so many of them already.

What is happening in India at the moment is that the technology is there, but not yet widely available. Not only the complexity of developing a transgenic plant means that many patented technologies may be required to produce a GM plant with the desired trait (Pary & Naseem, 2003:5), but laws and enforcement of intellectual property rights have not changed since 1972, when new chemicals, pharmaceuticals, and food and agricultural products were excluded from product patent protection (Pray & Basant, 2002: 48).

Introducing GURTs is one of the counter measures considered by the innovators to retain their share of monopoly profits, but it is taking valuable research time and resources away that could instead be dedicated to the priorities of the people and farmers living in India. In other words, as long as prices and restrictions to reproducibility remains high, and policy making bodies do not consider taking actions to reduce seed prices, very little of the benefits can be captured by the farmers, thus the innovators may gain today, but will lose potential income generating customers in the future if there is a backlash.

THE RELEVANCE OF APPROPRIABILITY

In India, not many firms inbred lines themselves except for the foreign suppliers. In turn, this raises a number of issues and affects the availability of crop variety to farmers as well as what local research laboratories can experiment with and then supply. It can be argued that by supplying only certain types of inbred lines, the combination of these would probably lead to a finite number of possible outcomes and hamper the creation of new varieties. That need not to be necessarily bad, but has consequences on the kind of work, projects and research that emerging nations can implement or carry out on their own.

Presently, about half of the seed sales are by public corporations (Pray & Basant: 2002) and innovative farming methods are normally first introduced by the government's specialised seed agencies (affiliated to the National Seed

Corporation), while the private sector takes over only when demand is consolidated. These firms then 'purchase' inbred lines from NARs (National Agricultural Research Centres) and develop new lines based on these. Assume that innovators in agribusiness invest time and resources in order to find cheaper, more efficient and better eco-friendly ways to increase agricultural production. When a breakthrough such as GM seeds is discovered, before commercialisation and distribution takes place, prices are set in such a way that given a certain expected demand, the costs of investment can be recovered as soon as possible in order to make a profit and reinvest. Competition from other innovators reduce the margin that it is possible to gain from the sale and prolong the time it takes to recover the initial investment.

Thus, unless the appropriate patent legislation granting temporary monopoly profits is in place the innovator delays the introduction of its discovery at great social cost. However, the above argument says little about what contributes to the diffusion of innovative technologies and portrays only a side of a wider debate. It is argued that if the benefits resulting from the introduction of an innovative technology can be captured evenly along a spectrum that goes from the innovators to the farmers then there might be better chances that the technology will spread.

In other words, that would be the degree of appropriability of the innovation. In the case of LDCs, one would probably expect that the farmers should be able to capture a considerable portion of the benefits. But what is it meant by benefits, and how are these measured? As the argument goes, with the appropriate legislation in place, the degree of appropriability can be defined as the income gained by each party involved (whether the innovators, the farmers or the distributors) after taking into account the price and income differentials of the new method as opposed to the old one, divided by the total income generated by the innovation.

According to Prof. Deodhar (2004) of the Indian Institute of Management, despite the essential regulation to secure appropriate standards for food quality and safety has been in place since the 1940s there are still too many eccentricities in the regulatory system that reduce their effectiveness and such legislation may also be ill-suited to deal with the issues that the introduction of biotechnology in agriculture raises. The usual suspects are found to be among the causes: corruption, unclear rules producing loopholes for the well connected and a Byzantine bureaucratic apparatus. For instance, offence and punishment are considered the same whether the verified adulteration is great or small and there is poor integration among the 6 ministries overseeing the process. This in turn affects the way in which the regulation and setting of acceptable standards of safety (such as the maximum acceptable intake of chemicals and other additives harmful to human health) are decided and the validity of some of the parameters taken into account to conduce experiments and orient public policy. At the time of our study news headlines were reporting the dangers of using 'acceptable standards', while de facto polluted, water in the soft drink industry.

Unsurprisingly, only recently there has been an effort to set up an institutional framework to deal with research and application of transgenic crops in India. Such attempt is based on the expectation of economic assistance from USAid and unfolds within the auspices of the Collaborative Agricultural Biotech Initiative (CABIO). As a result, institutions like the Department of Biotechnology (DBT), the Indian Council for Agricultural Research (ICAR), the National Centre for Plant Genome Research (NCPGR) and Indo-US Science and Technology Forum will jointly work together for setting up and steering the proposed institutional framework.

The reduction of import duties, together with privatisation of some government activities while increasing the use of effective public-private partnerships are among the objectives. However, as it is happening also in other parts of the world, large scale farming is pushing the use of GM to gain efficiencies in production, while smaller farmers are shifting food grains for cash crops to increase their income. This is also producing biases in crop research patterns away from what is considered necessary for nutrition (such as maize or corn) in favor of consumption (such as tobacco and sugar) and there are no serious biotech research investments in the five most important food crops in the semi-arid tropics: sorghum, pearl millet, pigeon pea, chickpea and groundnut (Pray & Naseem, 2003: 2).

POLICY IMPLICATIONS

As a member of WTO, India has opted for a sui generis intellectual property legislation stated in The Protection of Plant Variety and Farmers Act of 2001. According to this Act, process patents will be allowed on microbiological, biochemical and biotechnological processes (Sahai, 2004). This includes Plant Breeders Rights giving formal authority to licensed bodies to market, distribute, import or export a patented seed variety and severe punishments for non-compliers and illegal use of patented denominations. Interestingly the act recognizes the farmer not just as a cultivator, but also as a conserver of the agricultural gene pool (Sahai, 2003). In order to protect also the farmers' rights and avoid them being exploited by formal owners of breeders' right the scheme also envisages involving Non Government Organisation to guide illiterate farmers through the process. Unfortunately drafting of the scheme is still poor and thus prone to conflicting interpretations. For instance, Breeders Rights can only be given to local companies and are limited only to new plant varieties. However, also foreign companies should be treated equally.

The sides of the debate, which manifest themselves as we move from the farmers to the local institutional and legal framework to the international arena are many. The greatest risks seem to come from the introduction of mutually exclusive policies that could increase incoherence in the already uneasy framework governing the innovation and its various residual claimants. Likewise, not only our previous example Monsanto pateting Chapati raises important issues, but also as Devinder Sharma (Sharma, 2003) points out that something similar is happening with Bollgard. "Ironically", he says, "while cotton growers in the central region of the country find no buyers for their harvest, cotton imports are multiplying from 21,000 tonnes in 1999 to 49,000 tonnes in 2000. With the United States, China and European Union refusing to reduce their subsidies to cotton growers, there is no possibility for Indian farmers to find a footing in the international market".

Unfortunately, it seems that what raises greater concern within the current efforts to govern the innovation it is not how to make such technology as easily and cheaply available, but how to preserve the rents that it will generate to the innovators. But, GURTs and patents alone seem to be ill-suited to deal with the complexity of the issues surrounding the introduction of GM seeds in India. As an innovation, GM seeds hold great expectation and promises that are still to be fully exploited. Whereas the innovators justify their concerns under the grounds that the industry is capital intensive and that very large investments are required to enter it and support research in the area. The Green Revolution was introduced on the grounds of costs reduction and increased productivity, but has produced un-measurable social welfare losses considering the costs of health related diseases, let alone how it has affected the future productivity of the soil ad the purity of the water. Moreover, it has promoted a mentality that is hard to eradicate and the use of mono-culture for cash crops requiring massive amounts of pesticides and other chemical agents.

Therefore, new policies for the agricultural sector should consider not only how to increase in production and productivity, but also the possibility to increase farmers' (as well as others workers in the industry) available incomes together with better employment opportunities. For this reason public-private partnerships need to be assessed not only by looking at the overall increase and benefit of increasing R&D or seed sales, but also in terms of who (or which group of stakeholders) is able to capture the greatest amount of benefits by the introduction of this innovative technology, alias the degree of appropriability. In order for that to happen there will be the need to create better communication channels with technological institutions, forward linkages with markets and lateral linkages among the various groups that will be undertaking special licensing agreements and other public and private institutes.

Thus, patenting and enforcement of property rights legislation should create a hierarchy of ownership, cross-ownership, partnership arrangements and royalty payments that by all means aims to reduce market concentration and maximise diffusion of the new technology. In reality, the risks are high if legislastive measures will not be taken soon to ensure the former. Finally, future policies should aim at reducing the costs of making available less polluting and innovative farming methods by drafting a scheme which allows for a certain degree of appropriability to be kept by the farmers, and avoid them from being scavenged by formal Breeders Rights holders. In nuce both farmers and consumers should be able to make the choice by being informed about the origin of the product, its processes and the share of the price that each step of production will appropriate on the label of the end product.

CONCLUSIONS AND SUGGESTIONS

In conclusion, the introduction of modern biotechnology might not increase food production, but has the potential to reduce the impact of production methods associated with the (not so) Green Revolution on the environment. Our exploration of Indian's uneasy situation in agricultural markets and the emerging institutional panorama governing the technology shows that current research and trials are rather detached from the reality and needs of the farmers and the population at large and that in the absence of the appropriate legislation granting not only innovator' but also farmers' rights, the former risk taking extreme steps in patenting original life forms and traditional local food.

For the immediate future, the degree of appropriability will determine the possibility of diffusion of the new technology as well as its ultimate effects on development. That would need to go in tandem with developing seed varieties that are useful for farmers, which is not as obvious as it should be. Future research should include in the agenda the responsible use of biotech for the needs of farmers, customers and people in LDCs. For instance by identifying and breeding crop varieties resistant to biotic and abiotic stress, such as drought and salinity or by growing crops with embedded vaccinations that might be too expensive to get or too difficult to access in rural areas and by investing in food crops with a nutritional impact, other than only for consumption. Nevertheless, a thread of this report is extent to which modern biotechnology can be subsidised and for whom since Indian farmers cannot afford it. This leads in turn to two set of questions.

On one hand this leads to ask how can GM technology be disseminated to reach farmers at feasible prices. And how can indigenous varieties be protected by uncontrolled/unregulated patenting from outside players, which could furtherance the country's dependence on Western agribusiness. The Protection of Plant Variety and Farmers Act of 2001 and its more recent 2002 version are both good in principle, but if local Breeders Rights patents can be given only to new life forms without doing the same for foreign companies there is a clear risk of foreign appropriation of local varieties, effectively forbidding local companies to do the same.

In this process the rights of farmers and customers are poorly protected and the degree of appropriability uneven. The popular assumptions behind the introduction of GM in LDCs does not hold if lower regulatory barriers allow small scale production for seed exports in India, then growing the plants abroad and finally invading the LDCs' markets with GM crops imports as in the case of Bt Cotton.

On the other, there are a variety of concerns in relation to the legitimacy of GM foods themselves, not only on the economics of agricultural policy but for their unknown effects on human health, animals' life and the ecosystem. Arguably the area which encounters greater divergence of opinions, higher risk coefficients and uncertainties because of the lack of long-term studies on the effects of the use of such plant varieties for consumption by human beings. Both need to be considered highly on the agenda of regulators and scientists since GM foods are already present at various levels of the global food production chain. In order to achieve such objective there is a need to increase the availability of information to the farmer as well as the customer and to improve knowledge transmission systems between the various entities working on and with GM foods.

Such information should be made publicly available and then discussed at various levels of interaction among scientists, government entities, farmers and citizens. However, at the moment experiments on Public Participation and Governance of the Innovation NbE: see the update at the bottom of this post (11.11.04)] of the type endorsed by the FGB are unrealistic in a country like India where only a very small portion of the population has access to the Internet and possesses the available income to purchase a Personal Computer. By the same token, the two arguments are inter-linked in the very same essence of the problems and potentials related to the use of new technologies. Unfortunately, more seems to be invested in protecting the rights of the innovators rather than the needs of the farmers in India, thus a more balanced approach along the spectrum is needed.

That should not nevertheless divert our attention from the emerging institutional framework governing the innovation and allocating its property rights. It is strongly perceived that for the future an important objective should be to decrease prices, increase dissemination of the innovation and re-balance local and global policy inconsistencies.

Certainly, the considerations and suggestions advanced in this report remain limited to the agricultural sector, which need to be taken only as general given the size and diversity of a country like India.

Even if the focus throughout the report was to look at the emerging global/local dimension governing the innovation that should not be uncapped by other parallel reforms such as land reforms, the availability of modern methods of irrigation, improvements in the distribution infrastructure and communication mechanisms. To be sure, more tests will need to be undertaken as there is much unknown about the effects of GM seeds on the ecosystem and human health.

2

Science, Politics and Responsibility

Looking at the recent developments within one of the pillars of the welfare state (the health sector) it seems that, fours years after, the reach of these issues not only has become a locus of cogent political struggles in the United States, but also in Europe and in the developing world. The latest news regard a South Korean stem cell scientist, Hwang Woo Suk, who faked his research on stem cells (used for embryo cloning experiments). The various facets of the governance of scientific innovation, the role of experts and non-experts in such process and their responsibility in the translation of a powerful technology in innovation, have been the theme of a series of articles published on web-site of the Giannino Bassetti Foundation. These controversies in science and technology are increasingly framed as ethical and moral rather than as merely technical disputes, but then spark also grand political debates.

Governmentality and Innovation

The debate which divides politicians and scientists all over Europe extends beyond these considerations, which, looking at their immediate political impact, questions not just which source of evidence should be considered as legitimate science (the question of who politicians should trust), but also what should be the system of values to apply when deciding upon the policies deemed to be most appropriate to deliberate on the social impact of these technologies. Therefore, policies and regulations regarding the social impact of new technologies, which aim to minimise the sources of potential risks or dangers these may cause, have more than one domain of application, but are particularly sensitive areas because concern the life and death of millions of individuals. Biomedicine is but one of the untested territories in the field of science and technology studies that is propagating the feeling of uncertainty and indeterminacy that typically accompanies the introduction of a radical innovation. Especially because it is difficult to predict what the long term effects will be on human health, but also in the ecosystem of our planet. Think also how the concerns regarding the validity of research being expressed here for the field of biomedicine may also apply for the fields of biotech, genetics, nanotech, and technologies of information and communication.

The possibility to define a role for science and indeed also of scientists in issues of interest for the well-being of society has shifted the focus of public opinion only on the external representation of a limited understanding of the scientific domain. Take the example of a pharmaceutical company filing for the approval of a new drug, who would doubt about the importance of applying a scientific rationality (and indeed great precaution) before bringing the drug to the attentions of the regulators, and only when it has been tested and considered safe? Patti Lather (2004) reminds us that in Focauldian terms, policy is one of three technologies of governmentality, the others being diplomatic/military and economic. Governmentality is associated to the concept of biopolitics, where the state intervenes at a distance (i.e. via regulation) in the striking attempt to balance such intervention with a liberal approach to politics. Such 'political arithmetic' (Foucault, 1998) makes particular kinds of discourse both possible and necessary.

The debate regarding the responsibility and involvement of scientists in such decisions (or better said those scientists that fail to abide to the principles of Science with the capital 'S') probably would be hard to explain to historians that wrote about the Enlightenment (a period marked by scientific and technological discovery) after the Dark Ages. Since then science has been a power engine of economic development, advancing the status of human knowledge and capacity contributing to innovations that have raised the living standards of the population. The list of inventions that have transformed or revolutionised people's everyday lives is endless. One needs just to look at the office space today as opposed to ten years ago to start noticing some remarkable differences. And of course this is only part of a longer history of continuous improvement upon pre-existing discoveries.

The artifacts of innovation surround us when we want to call our relatives or friends, book a flight or report a fault in one of our many gadgets. Such artifacts are, however, only the effect of a far more complex process of negotiation and confrontation between a diverse set of actors responsible for (or at least involved in) various degrees with their conceptualisation, design, development and production. Such actors belong to, as well as are part of, different and sometimes competing domains. From a Foucaldian perspective, what becomes of interest to disentangle their interaction is how such distinctions are created ('good' vs. 'bad'; 'responsible' vs. 'irresponsible') and upon which bases of knowledge do they build their authority. For simplicity, we can outline their agendas as related to Science, Politics and Technology.

Science, Politics and Technology

Distinguishing between Science, Politics and Technology leads to a different level of understanding the role between Science (with the capital 'S') and the politics of science, a particular area of scientific activity, which does not reflect the ideals upon which its aspiration is enacted within the fragmented existence of competing and conflicting communities. Beyond the ideal-Science-we find the existence of a 'varieties of science' hypothesis. In this context Politics is important because it defines who is important. However, Politics is not the same as 'policy', a more ambiguous term which can be used to refer to:

- Government, administration, the conduct of public affairs; political science;
- Political sagacity; prudence, skill, or consideration of expediency in the conduct of public affairs; statecraft, diplomacy; in bad sense, political cunning;
- A course of action adopted and pursued by a government, party, ruler, statesman, etc.; any course of action adopted as advantageous or expedient.

At this stage is thus important to make a further distinction between Science and Technology and especially their underlying objectives. By those who practice it, science can be considered a disinterested search for the pleasure of discovery, with the ambition to advance the status of human knowledge against the backdrop of ignorance which surrounds the sphere of knowledge of the laymen. Technology instead can be understood as expressing a different rationality, which is guided by an entirely different set of interests. In the Heideggerian conception, Technology can be interpreted as the construct of science, 'a power whose great role in determining history can hardly be over estimated' (Heidegger, 1976). Thus, when a reference to science is made, perhaps many consider 'scientific' the latter, not the former understanding which are sometimes confused in general discussions about this topic.

Unfortunately, it is not only a matter of academic debate because policies and regulations are based on definitions and are implemented as understood also by the (political) meaning which is assigned to them. A fascinating example of this case is made in an article regarding the level of safety of microwaves produced by mobile telephones. Interestingly, however tight or loose the definition of the level of acceptable exposure to the microwaves produced by mobile telephones, this does not seem to stop anyone from buying mobile telephones or reduce the growth rate of this sector, despite in some countries it already is a nearly saturated market. Despite the controversies, mobile phone sales worldwide have increased by nearly 150 million units sold in 2005 (21% more than the units sold in 2004) and the higher percentages of growth rates can be found in the fast developing areas of Latina America, Far and Middle East.

So there are indeed many ways to interpret the contours of the complex relationship existing in the space of interaction of science, politics and society. According to Richard A. Pielke (2002) a mutually reinforcing 'iron triangle' theory explains why science has become political. The theory can be briefly explained as follows: in a corner are the politicians, always careful about the opinions of their voters and generally seek to avoid the consequences of having to make a decision; in another are the scientists, which are empowered to provide policy answers with the funding they receive for their research; and in a third corner are interest groups and lobbies, which look for scientific evidence 'to provide a compelling justification for their political, societal, environmental or business goal'.

Both precaution and responsibility apply to all of the groups identified by Pielke's colorful representation.

A British Perspective on the Governance of Science

On the other hand, moving to another set of issues, these ideas do not necessarily apply to the whole of the scientific profession, and the generally accepted methodological prescriptions that can hold in a stable and predictable environment need not to apply when transitional activities are involved such as the definition of responsibilities in a new field of study or discipline. Acknowledging that also research is a human activity therefore means to understand first of all that researchers are human and therefore bring into the object of their investigation also their tacit knowledge, emotions, moral and political convictions that cannot immediately be rationalised in methodological prescriptions. However, there are still relevant sources of risk that need to be taken into account both when evaluating as well as when producing research:

> *The risk of producing misleading information, previously mentioned in the Bassetti Foundation web-site by the name of the principle of precaution. This includes also the conflict of interests that could arise in conjunction with the definition of issues and terms may arise especially when setting regulations concerned with the level and standards of health and safety. 2. The potentially restrictive effects of method on criticality, alias the first step towards a responsible choice on behalf of the researcher to produce an account which is 'truth'. A responsibility towards putting aside a predictable set of values, beliefs and ideologies and engage with an unknown world.*

We discussed some of these highly controversial ideas with Prof. Nikolas Rose director of BIOS, a multidisciplinary research centre analysing the practice and implications of developments in bioscience, biomedicine, biotechnology and society at the London School of Economics. As he elegantly put it:

> *'What biotechnology, genetics and technologies of information and communication have in common is that have been introduced without the possibility to*

predict what the medium to long term implications would be. No one though that these innovations would have advanced humanity and make available new sources of knowledge in the way the Internet did. However, no one thought that it would also help pornographers or be used to make public the sequencing of the human genome'.

And suggests that: 'For instance I am my collaborators at BIOS are researching on the economic drivers underlying the development of some of these technologies. Last year we explored the role of enhancement technology, this year we are looking at biotechnology and biocapital. The extent to which many of the developments in this field are being driven by tight speculative investments are at the centre of our discussions. One can think of the intersection between, on one hand, the political, and on the other the powerful economic drivers, but there is a demand side as well. 10 years ago the idea of having a kidney transplant was a rather unusual intervention, but now that it is possible, there is a general view that there is a world shortage of kidneys. Everybody almost feel they have a right to a kidney 'on demand'.

And because there is a technological innovation that has made it possible, it also created a market and the general idea that the body is a set of replaceable parts. Therefore, we need to break down the areas of impact of biotechnologies, such as the use of biotechnology in industrial processes, biomedicine or the agricultural sector. Each of these areas poses new questions of interest about what should be considered legitimate science, not just any science, but science with the capital 'S'. What I would stress is that the governance of science influences the future very heavily and it is not surprising that every European government has set up huge task forces to work out the effects of the introduction of these technologies. There is clearly some wish, some endeavor to govern these technologies and the future of science.'

The Challenges Ahead

One perhaps should also reflect about what the challenges ahead will be, especially for those countries which do not invest enough in education and research and in how to improve the mechanisms of ethical accountability at the nexus between science and politics. Such responsibility involves improving the goals of the governance of technological innovation. As a consequence, one may ask how could it be possible to advance the objectives of the Lisbon strategy, which envisioned a raise of the resources available to science and education capping a 4-5% of the budget of the European Union? And especially if the ambitious objective of becoming one of the most progressive societies based on knowledge is to be reached?

In this context one of the greatest challenges consists in developing the 'know how' necessary for discovery to become innovation, which requires a network of expertise able to produce new cohesive forms of knowledge, alias its actualisation which generally results from the intertwining of discovery with actuating power (Bassetti, 2003). But because the framework of reference is not any longer merely national, but increasingly transnational, the forces that need to be combined in order to achieve that are based, on one hand, on the development of an educational system open to learning and on the a skilful combination of national, regional and international expertise.

Hence, the relevant question here is 'how can science be part of the political process and yet separate from its Technology? A recently published book titled 'Nature's Experts: Science, Politics, and the Environment' by Stephen Bocking (2004) addressed a somehow similar question. Written by an environmental scientist, it addresses a theme of interest to all those that have a stake in the debate about the history of science and its future delineation. He explains an important point with an insightful quote from Dorothy Nelkin (which we reproduce below):

> *"As scientists debate the various sides of political issues, their involvement undermines the assumptions about the objectivity of science, and these are precisely the assumptions that have given experts their power as the neutral arbiter of truth".*

However, something is missing from both Pielke's triangle and Booking's account, something that can the triangle a square: citizens and consumers living in the emerging knowledge society, which are redefining the contours of the governance, risk acceptance and responsibility of the processes of governance of scientific and technological innovations.

Over the Internet, the number and of sources available to evaluate the credibility of scientific research, its health risks and political impact has increased exponentially in recent years. Suggesting the need of a new maturity and a new sense of accountability on the side of anyone who becomes involved in the process of decision making that will affect the choices of civil society. Perhaps another danger resulting from the way in which the problem of science, politics and responsibility is framed with regards to the governance of innovation leads to the understanding that it can be fixed merely as administrative issue.

Increased awareness and capacity on the side of previously excluded sections of society, are now calling for a new and overt role in the way in which previously closed and relatively unaccountable decisions were taken. The opportunity of greater participation in the context of the knowledge society reinforces the need to focus our attention on the constitution of appropriate mechanisms that can promote a new ethic of (political) responsibility in the way in which decisions affecting the sphere of human life and health may be taken. This understanding of the complex relationship animating the 'be sidedness' (to use the expression suggested by Mario Castellaneta, a reader of this Blog) between science and society is indicative of an emergent new paradigm of knowledge production and validation.

In such process, science is not anymore closed and unaccountable (the ivory tower referred to by Bassetti in his article on 'New Science and New Politics'). Having said that and with reference to the issues that have been addressed by the previous articles, is thus necessary to make an important distinction between Science, Technology and Politics and their underlying objectives without generalising the occurrence of exceptions.

However, in the traditional duality expressed by the principle of precaution and the principle of responsibility my fear is that, as Chrisanthi Avgerou of the London School of Economics puts it, the more we are preoccupied with perfecting particular principles of research legitimacy, the more we stifle the possibility for critical debate across diverse socio-cultural settings and disciplines. In other words, the reason why such perspective may be of distinct quality that it is not appropriate for advancing knowledge in society, but that they advance too well knowledge that serves unquestioned social ends that may be of dubious political and moral status.

Richard Feynman, Noble prize winner in 1956 for quantum physics and a pioneer in nanotechnology, supported the idea that it was absolutely coherent to be uncertain and scientific at the same time.

The freedom of doubt is what he considered to be one of the greatest achievements of science and in that responsibility towards society he saw a close relationship with the fundamental principle of democracy. Nevertheless, this would need to take into account also the new security environment after the terrorist attacks both in New York and in London and the necessity to balance the appropriate provision of public support for research (an administrative matter) without transforming at the same time such openness and the public availability of results, into weapons which may lead to catastrophic acts of terrorism (a do-it-yourself nuclear weapon).

However difficult and unpopular this may prove to achieve, one should nevertheless consider the fruits that can result form having scientific knowledge available in the public domain. Helga Nowotny, chairwoman of the European Research Advisory Board of the European Commission says that such new paradigm is 'socially distributed, application-oriented, trans-disciplinary and subject to multiple accountabilities'. This is a timely issue to discuss as one of the challenges ahead, considering the imminent launch of the new European Framework Programme.

CONCLUSION

Investment in science remains an important prerequisite to promote innovation. However, innovation brings into play new forms of dangers or risks that civil society should debate openly.

Balancing the views, interests and values of all the participants to innovation is what legitimises also the knowledge required to influence its path. Acknowledging these dynamics leads us to suggest the inclusion of a certain element of criticality in future policy initiatives. Giving immediate priority to frame the discussion in the political domain and to find a political answer that is sometimes escaped.

If politics is considered as the locus of reconciliation of ethical and economic dilemmas, which resolves the dispute in terms of equity rather than efficiency it is not necessarily true that involvement of science means to look only or primarily upon issues of administrative efficiency. Should the decision of the right of the level of minimal health care assistance be a political decision or a scientific one? Should in this case science be called to express an opinion, or a judgement? Will this be enough to avoid a political struggle between such two different rationalities? However, ethical and political studies are still 'scientific', although not necessarily in the term implicit in the realm of technocracy. The latter can be considered as an attempt to rationalise and deliberate on issues which cannot be rationalised without an increasing sense of responsibility on the side of those who have the responsibility to govern innovation.

In conclusion, we have tried to shed some light in the processes animating the debate about science, politics, responsibility and the governance of technological innovation. A quote from Giuseppe Longo seems appropriate to conclude this essay, hoping that it will be the beginning of a constructive debate: 'in order to populate the market the technology does not wait any longer for science and its patents of legitimacy'. It is also worthwhile reminding that the universal social value of science and research consists also in its reflexivity.

An inward look to the process of discovery itself and not just an investigation of the world of innovation and against the universally preconceived values of method, is essential in what may be considered good science and good research. In doing so we have highlighted that in nearly all circumstances the way in which the term science is used is misleading without a proper definition and therefore it is perhaps appropriate not to define the debate in terms of governance of science or of politics, since that is not what reflects the responsibility of such governance according to the Heideggerian definition of Technology.

3

BIOTECHNOLOGY DEVELOPMENT STRATEGY

INTRODUCTION

Biotechnology, globally recognized as a rapidly emerging and far-reaching technology, is aptly described as the "technology of hope" for its promising of food, health and environmental sustainability. The recent and continuing advances in life sciences clearly unfold a scenario energized and driven by the new tools of biotechnology. There are a large number of therapeutic biotech drugs and vaccines that are currently being marketed, accounting for a US$40 billion market and benefiting over a hundred million people worldwide. Hundreds more are in clinical development. In addition to these there are a large number of agri-biotech and industrial biotech products that have enormously helped mankind. The Indian Biotechnology sector is gaining global visibility and is being tracked for emerging investment opportunities. Human capital is perceived to be the key driver for global competitiveness. Added to this is a decreasing appetite for risk capital in developed countries, which has led to a decline in the biotechnology sector in these regions where survival lifelines are being provided by the lower cost research environs of the developing world such as India. For a country like India, biotechnology is a powerful enabling technology that can revolutionize agriculture, healthcare, industrial processing and environmental sustainability.

The Indian biotechnology sector has, over the last two decades, taken shape through a number of scattered and sporadic academic and industrial initiatives. The time is now ripe to integrate these efforts through a pragmatic National Biotechnology Development Strategy. It is imperative that the principal architects of this sector along with other key stakeholders play a concerted role in formulating such a strategy to ensure that we not only build on the existing platform but expand the base to create global leadership in biotechnology by unleashing the full potential of all that India has to offer.

IMPORTANCE OF BIOTECHNOLOGY

Biotechnology can deliver the next wave of technological change that can be as radical and even more pervasive than that brought about by IT. Employment generation, intellectual wealth creation, expanding entrepreneurial opportunities, augmenting industrial growth are a few of the compelling factors that warrant a focused approach for this sector.

Vision and Mission: Biotechnology as a business segment for India has the potential of generating revenues to the tune of US$ 5 Billion and creating one million jobs by 2010 through products and services. This can propel India into a significant position in the global biotech sweepstakes. Biopharmaceuticals alone have the potential to be a US$ 2 billion market opportunity largely driven by vaccines and bio-generics. Clinical development services can generate in excess of US$1.5 billion whilst bioservices or outsourced research services can garner a market of US$1 billion over this time scale. The balance US$500 million is attributable to agricultural and industrial biotechnology.

India has many assets in its strong pool of scientist and engineers, vast institutional network and cost effective manufacturing. There are over a hundred National Research Laboratories employing thousands of scientists. There are more than 300 college level educational and training institutes across the country offering degrees and diplomas in biotechnology, bio-informatics and the biological sciences, producing nearly 500,000 students on an annual basis. More than 100 medical colleges add ~17,000 medical practitioners per year. About 300,000 postgraduates and 1500 PhDs qualify in biosciences and engineering each year. These resources need to be effectively marshaled, championed and synergized to create a productive enterprise. India is reorganized as a mega bio-diversity country and biotechnology offers opportunities to convert our biological resources into economic wealth and employment opportunities. Innovative products and services that draw on renewable resources bring greater efficiency into industrial processes, check environmental degradation and deliver a more bio-based economy.

Indian agriculture faces the formidable challenge of having to produce more farm commodities for our growing human and livestock population from diminishing per capita arable land and water resources. Biotechnology has the potential to overcome this challenge to ensure the livelihood security of 110 million farming families in our country.

The advancement of biotech as a successful industry confronts many challenges related to research and development, creation of investment capital, technology transfer and technology absorption, patentability and intellectual property, affordability in pricing, regulatory issues and public confidence. Central to this are two key factors: affordability and accessibility to the products of biotechnology. Policies that foster a balance between sustaining innovation and facilitating technology diffusion need to be put in place.

There are several social concerns that need to be addressed in order to propel the emergence of biotechnology innovation in our country such as conserving bioresources and ensuring safety of products and processes. Government and industry have to play a dual role to advance the benefits of modern biotechnology while at the same time educate and protect the interests of the public. Wide utilization of new technologies would require clear demonstration of the new added value to all stakeholders.

The National Science and Technology Policy of the Government and the Vision Statement on Biotechnology issued by the Department of Biotechnology have directed notable interventions in the public and private sectors to foster life sciences and biotechnology. There has been substantial progress in terms of support for R&D, human resource generation and infrastructure development over the past decade. With the introduction of the product patent regime it is imperative to achieve higher levels of innovation in order to be globally competitive. The challenge now is to join the global biotech league. This will require larger investments and an effective functioning of the innovation pathway. Capturing new opportunities and the potential economic, environmental, health and social benefits will challenge government policy, public awareness, educational, scientific, technological, legal and institutional framework.

The issue of access to the products arising from biotechnology research in both medicine and agriculture is of paramount importance. Therefore, there should be adequate support for public good research designed to reach the unreached in terms of technology empowerment. Both "public good" and "for profit" research should become mutually reinforcing. Public institutions and industry both have an important role in the process.

The National Biotechnology Development Strategy takes stock of what has been accomplished and provides a framework for the future within which strategies and specific actions to promote biotechnology can be taken. The policy framework is a result of wide consultation with stakeholders-scientists, educationists, regulators, representatives of society and others and reflects their consensus. It focuses on cross-cutting issues such as human resource development academic and industry interface, infrastructure development, lab and manufacturing, promotion of industry and trade, biotechnology parks and incubators, regulatory mechanisms, public education and awareness building. This policy also aims to chalk out the path of progress in sectors such as agriculture and food biotechnology, industrial biotechnology, therapeutic and medical biotechnology, regenerative and genomic medicine, diagnostic biotechnology, bio-engineering, nano-biotechnology, bio-informatics and IT enabled biotechnology, clinical biotechnology, manufacturing & bio-processing, research services, bio-resources, environment and intellectual property & patent law.

Several state governments have enunciated biotech policies spelling out a comprehensive blueprint for the sector. It is, therefore, prudent to have a National Biotech Development Strategy that charts an integrated 10-year road map with clear directions and destinations. This is the time for investment in frontier technologies such as biotechnology. It is envisaged that clearly thought-out strategies will provide direction and enable action by various stakeholders to achieve the full potential of this exciting field for the social and economic well being of the nation.

KEY POLICY RECOMMENDATIONS AND INTERVENTIONS

Biotechnology is a knowledge-driven technology, which needs to be driven by a flow of new ideas and concepts in the development of new tools for research, new processes for manufacturing and innovative business models. Rapid responses are needed to meet the challenges as they unfold and there is a requirement for specialized personnel and centres of excellence for R&D. The policy goal for the next decade is to facilitate the availability of scientific and technical human resource in all disciplines relevant to the life science and biotechnology sector. In order to build a successful biotechnology sector, large talent pools are required in multiple scientific disciplines such as molecular and cell biology, chemistry, physics, engineering, bioinformatics, medicine, agriculture, microbiology, technology transfer & commercialization, bioentreprise & biofinancing and intellectual property rights management. Product and process development are inter-disciplinary in nature and deficiencies in specific areas may weaken the whole sector. The key issue is the manner in which to create an effective interface across disciplines.

Reliable estimates of human resource availability for the next 10 years are required. Expert consensus indicates that there is adequate enrollment currently at the post-graduate and under-graduate levels, however the quality is inconsistent. Areas such as intellectual property rights, regulatory issues and industrial training have received inadequate attention. There is a consensus that there is an urgent need to augment the number of Ph.D. programs in the Life Sciences and biotechnology. A strong pool of academic leaders is key to sustained innovation.

Strategic Actions:

(i) National Task Force on Education & Training

- A National Task Force will be created to formulate model undergraduate and postgraduate curricula in Life Sciences keeping in view, future needs. The said curricula must address the underlying need for multi-disciplinary and inter-disciplinary learning and the appropriate stage for biotechnology training.

(ii) Need Assessment

- There would be need assessment in 2005 for the next five years and close monitoring during the period for interim changes.
- A 10-year perspective plan for human resource will be prepared every five years.

(iii) Curriculum Development

- Course curricula will be reviewed and improved in consultation with industry and research establishments and standard e-learning modules will be developed for specific skill areas such as IPR, regulations, and bioentreprise.
- Hands on exposure to M.Sc. biotechnology students will be enhanced through an extended industry internship as well as through short-term placements at CSIR and other appropriate National Institutes.
- Dual degree programs in biotechnology that include regulatory matters, IPR and bio-enterprise management will be encouraged and supported by the Department of Biotechnology.
- Emphasis will be given on training of high quality technicians and technologists in skills required by the industry by establishing Regional training centres at diploma, graduate and postgraduate levels.

(iv) Quality Improvement

- An accreditation mechanism will be put in place for ensuring minimum standard of education and training at the post graduate and undergraduate levels. Base requirements for teaching and laboratory infrastructure will be specified and enforced.
- Teachers training programs will be taken up by creating regional teachers training centres

(v) Strengthening of Teaching and R&D in Life Sciences and Biotechnology in the University System

- Strengthening R&D in Life sciences and biotechnology in the university system will be accorded high priority. This is considered important for improving the quality of education and providing exposure to new technologies for students at various levels. Specific mechanisms to achieve the goal will include
- Creation of inter-disciplinary centres of excellence with world class infrastructure in key areas
- Program support to encourage inter-departmental networking
- Visiting professorship and creation of industry sponsored chairs in partnership with the Department of Biotechnology.

(vi) Attracting Talent to Life Science and Biotechnology

- Bright students will be attracted to take up careers in biology and biotechnology through special scholarships. Summer assignments at Academic and Industry research laboratories will be introduced at the school level to create interest in the fields of biotechnology and biology
- Women scientists will be encouraged to take up careers in biotechnology. Service conditions will be liberalized for women to be able to return to research/academics after maternity breaks.

(vii) Creating Science & Technology Leaders for the Industry

- The number of Ph.D fellowships offered by the Department of Biotechnology will be increased to 200 per annum
- Public-private partnerships will be encouraged in Ph.D programs through creation of the 'Bio-edu-Grid'-a network of universities and industries facilitating pooling of resources.
- Masters degree level professionals in industry will be encouraged to undertake Ph.D. programs while

retaining their jobs through industry-university tie-ups.

(viii) Arresting and Reversing Brain Drain

- As mentioned earlier, the number of postdoctoral fellowships offered by the Department of Biotechnology will be increased to 200 per annum in order to attract talent.
- Outstanding young investigator grants in biotechnology will be introduced. This will provide a package including salary support, research grant, equipment and opportunities to attend national and international conferences. The salary support under the scheme will be at par with that of entry-level faculty positions.
- Information on availability of positions in education/ research establishments and industries will be provided on a website to facilitate employment of scientists with specific skills at appropriate positions.
- A database of scientists working in different areas of biotechnology within and outside the country will be created to utilize the expertise appropriately.

(ix) Enabling Working Conditions for Scientists to Undertake Industry Oriented Research

- Lateral mobility of scientific personnel: Scientists working at universities and research institutions may be allowed to work in industries for commercialization of their research efforts. This could be in the form of secondment or consultancy with industry or by a sabbatical for three years during the working life of scientists
- Dual/adjunct faculty positions: Researchers working in university/research institutions may be allowed to hold positions in the industry and vice-versa
- Joint salary support: Faculty employed in academic institutions may be allowed to hold positions for a period of time in which their salary is contributed

both by the industry and the academic institution on a mutually agreed basis. (Such an arrangement will work well only if the teaching requirements of the academic institutions are made obligatory).

- Rapid travel grants: Rapid travel grants scheme for approval within two weeks for young scientists to interact with mentors and industry collaborators would be initiated.
- Institute Innovation grants through the Department of Biotechnology to fund academic researchers to develop their concepts into patentable and more importantly licensable technologies. Such grants may be utilized for the purpose of providing additional infrastructure and manpower, patenting costs as well as costs related to proof of concept studies.

These steps will ensure that the large available resources of human talent in biotechnology are supported and this will guarantee the progress of the biotech sector.

INFRASTRUCTURE DEVELOPMENT AND MANUFACTURING

The strength of a biotechnology company lies in up scaling a number of proven technologies-diagnostics, vaccines, products, and processes-for fine-tuning and large-scale production. While Indian industry is strong in product development and marketing for commercial benefits, biotechnology in India still lacks the infrastructure required to take up R&D in areas like molecular modeling, protein engineering, drug designing, immunological studies, pre-clinical studies, clinical trials, etc.

In order to get the best from public and privately funded biological/biotechnological research, it is imperative to utilize the infrastructure generated optimally for societal benefits. The concept of contract research organizations (CROs), contract manufacturing organizations (CMOs), contract packagers, lab services providers etc., is steadily taking shape in India.

In the area of biomanufacturing, industry estimates that the market for biogenerics in India is expected to see a 43 percent jump from Rs 308.50 crore in 2001 to Rs 1,305.7 crore in 2005 and projected to reach Rs 1,864.3 crore by 2007 registering a growth of 19 percent. India's strengths lie in the availability of educated and skilled manpower, proficiency in English, low capital and operational costs and the proven track record in meeting international standards of quality. There is ample proof that the Indian companies are committed to global standards. According to reports, outside of the US, India ranks the highest with 61 USFDA-approved plants and in excess of 200 GMP certified pharmaceutical manufacturing facilities.

Strategic Actions:

(i) Department of Biotechnology will act to facilitate a Single Window Clearance mechanism for establishing Biotechnology plants.

(ii) Encourage private participation in infrastructure development like roads, water supply and effluent treatment.

(iii) Depositories of biological materials will be created in partnership with industry on IDA model for agriculturally important organisms, medically important organisms, plasmids, cosmids and constructs of special nature generated with adequate human interventions

(iv) State of the art large animal house facilities with GLP will be created for testing candidate vaccines and biotherapeutics. Testing facilities will be created for GMO/LMO

PROMOTION OF INDUSTRY AND TRADE

The emergence of India as a global player in the biotech sector requires government to play the role of a champion and foster an international competitive environment for investment and enterprise development. India's strategy must be to get more value from its R&D investment and from IPR generation.

The Biotechnology sector has in recent years witnessed accelerated growth. With approximately 200 industries the growth of the biotech sector in India has been rapid. Current estimates indicate that the industry grew by 39% annually to reach a value of US$ 705 million in 2003-2004. Total investment also increased in 2003-2004 by 26% to reach US$ 137 million.

Exports presently account for 56% of revenue. Currently the biopharma sector occupies the largest market share of 76% followed by bio-agri 8.42%, bioservices 7.70%, industrial products 5.50% and bioinformatics 2.45%. The bioservice sector registered the highest growth (100%) in 2003-2004 with bioagri 63.64% and biopharma 38.55%.

The current policy review envisages an annual turnover of US$ 5 billion by 2010. India has to develop its own biotechnological and pharmaceutical products to ensure quality and affordability for global trade. In addition to opportunities in drug discovery and development there are significant openings to provide services to the worldwide biotech and pharmaceutical industries and to leverage low cost high quality manufacturing with a global discovery potential. Capitalizing on these opportunities would create many new valuable jobs in India as we have seen in the outsourcing and service industry.

However, to achieve the targeted business volume, several new challenges have to be met. These are predictable and enabling policies, increased public and private support for early or proof-of-concept stage of product development; improved communication among stakeholders in the sector, public-private partnerships, integration of the Indian biotech sector globally; and improved infrastructure. The vision is to maximize opportunities in the area of contract research, manufacturing and to promote discovery and innovation.

The Biotech industry being capital intensive in nature has historically relied on venture capital from public and private sources. India needs to provide active support through incubator funds, seed funds and provision of various incentives in order to develop the biotech sector.

In a highly competitive and fast moving business environment, innovative capacity is an important determinant of the ability to create a continuing pipeline of new products and processes. Innovation covers knowledge creation (R&D), knowledge diffusion (education and training) and knowledge application (commercialization). Innovation is not a one-time event; instead, it has to continuously respond to changing circumstances for creating sustainable growth. Innovation is measured in terms of external/domestic patent applications;

human capital devoted to R&D, government expenditure on R&D proportionate to country's GDP, business funded expenditure on R&D, indigenous technologies standardized, demonstrated and transferred to industry for commercialization; and the number of spin off companies created. Clear government policies for promotion of innovation and commercialization of knowledge will propel the growth of the biotechnology sector.

Strategic Actions:

(i) *Innovation:*

- Basic and translational research in key biological processes and new materials will be supported as innovation for tomorrow. Access to the knowledge generated will be improved by supporting knowledge and social networks among stakeholders so that those with appropriate skills can convert the research output into useful products and processes.
- Research to promote innovation must be supported increasingly on a cooperative rather than a competitive basis. This requires effective communication among science agencies, research institutions, academia and industry.
- To promote India as a hub of innovation, a network of relevant stakeholders should be developed. Public investment should be used as a catalyst to promote such clustering and networking as this can lead to enhanced creativity by sharing of expertise, resources and infrastructure.
- Availability of human resource would be ensured at each phase of the product cycle.
- Strengthening technology transfer capacity

It is proposed to create several national/regional technology transfer cells (TTC's) over the next 5 years to provide high caliber, specialized and comprehensive technology transfer services. The services would include: evaluating technology and identifying potential commercial uses, developing and executing and intellectual property protection strategies identifying potential licensees and negotiating licenses. Each TTC would service a cluster of institutions in a region or a large city. Optimal delivery of services by the TTCs requires professionals with background in industry and science, wide networks, an external focus and high level licensing skills. The best practices for effective technology transfer will be benchmarked.

The skills of existing technology transfer professionals will be upgraded by a combination of specialized training courses, including linking to important programs redesigning the incentives and career paths for posting.

Scientists and other innovators shall be equipped with a better understanding of markets and commercialization pathways, the process of technology transfer, the strategy of protecting intellectual property rights and industrial licensing.

(ii) *Fiscal and Trade Policy Initiatives:* Biotechnology firms are by far the most research intensive among major industries. On an average the biotechnology sector invests 20-30 % of its operating costs in R&D or technology outsourcing. Government support, fiscal incentives and tax benefits are therefore critical to this sector. These measures will also help to capitalize on the inherent cost effectiveness of the Indian biotech enterprise. The suggested interventions include:

- Exemption of import duties on key R&D, contract manufacturing/clinical trial equipment and duty credit for R&D consumer goods to enable small and medium entrepreneurs to reduce the high capital cost of conducting research.
- Extending the 150 % weighted average tax deduction on R&D expenditure under section 35 (2AB) until 2010 and to permit international patenting costs under this provision and enable eligibility of expenditure incurred with regard to filing patents outside India for weighted deductions u/s 35 (2 ab)

- Enable lending by banks to biotech companies as priority sector lending. Currently banks are almost averse to lending to young biotech companies. In order to encourage banks to lend and provide banking services to the biotech sector, a significant push through appropriate policy guidelines from the Reserve Bank of India is necessary. Currently lending to agri-businesses as well as investment in Venture Funds by banks is categorized as Priority Sector Lending. Biotech as a business has similar characteristics in terms of risk as well as gestation time lines and it is therefore recommended that lending to Biotech be also categorized as Priority Sector lending.
- Remove customs duty on raw materials imported into India, where the finished product is imported duty free. Life Saving Drugs imported and sold in India are exempted from paying customs duty; whereas raw materials for diagnostics and other pharmaceutical biotech products manufactured in India are levied customs duty. To promote the indigenous manufacturing industry and make it competitive globally, raw materials imported by Indian manufacturers should be eligible for Duty Drawback.
- Rationalization of import and export of biological material is considered critical for clinical research and business process outsourcing.
- Simplification and streamlining of procedures for import, clearance and storage of biologicals, land acquisition, obtaining environmental and pollution control approvals would be simplified and streamlined within shorter time frame lines through consultations with various central and state government departments.
- As an effective regulatory mechanism has been put in place though recent interventions, Foreign Investment Promotion Board (FIPB) approval for equity investment may no longer be necessary.

- Joint R&D collaboration and generation of joint IP though global partnerships would be fostered.
- International trade opportunities would be promoted to guide R&D investment Indian biotech strengths would be aggressive by promoted globally.
- Efforts would be made to remove hurdles for contract research especially for input output norms and tax on revenue generated through contract research/ R&D.
- Easy access to information, regarding legislation and rules and regulations for transboundary movements of biologicals would be promoted.
- Current standards and safety of products would be enhanced.
- Efforts would be strengthened to promote acceptance of Indian regulatory data internationally.
- Research, trade and industrial partnership would be fostered at regional and sub-regional levels.
- A "cluster" approach would be encouraged to operations. One significant features of the industry is the fluidity and variety of its inter-company relationships, traditionally much greater than in other industries. It has relied to a considerable degree on contracting and outsourcing, especially "upstream" in R&D through various licensing arrangements and "downstream" through co-marketing agreements.
- Collaborative knowledge networks would be promoted. Expanded sharing of information, including creation/use of collaborative knowledge networks (CKN), can greatly enhance a company's performance under a cluster approach. Managing the many external relationships is complex. Flexible and pervasive communications systems that allow information to flow effortlessly within and between contracting organizations will provide the key to success. Increasingly, IT advances, including web-

based approaches, will provide the foundation for these systems.

(iii) *Public Investment for Promotion of Innovation and Knowledge Commercialization:* Availability of financial support for early phase of product development to establish proof-of-principle is the key to sustaining innovation. In this context, it is proposed to institute 'Small Business Innovation Research Initiative' (SBIRI) scheme through the Department of Biotechnology in 2005-06 for supporting small and medium size enterprises as a grant/loan. Companies with up to 1000 employees will be eligible. The scheme will support pre-proof of concept, early stage innovative research and provide mentorship and problem solving support in addition to the grant/soft loan. The SBIRI scheme will operate in two phases of innovation and product development.

- *SBIRI Phase-I:* The funding in this stage will be provided for highly innovative, early stage, pre-proof-of-concept research. Preference will be given to proposals that address important national needs. The maximum amount of funding to an enterprise will be limited to Rs. 50 lakh with not more than 50% of it going as grant and the remaining as an interest free loan. For projects to be considered at this stage, though a partner from a public R&D institution would be considered important, it will not be a mandatory requirement for those companies that have good quality scientists. This should encourage high quality scientists to agree to work in small and medium biotech companies, a change from our traditions. The R & D requirement of the public institution will be met through a grant.
- *SBIRI Phase-II:* It is expected that some of the proposals funded with SBIRI Phase-I will establish the proof-of-concept. At this stage, the ability of the project to get venture capital funding improves. Such projects will be eligible for Phase-II funding.

Some projects could be eligible for direct phase-II support. It is proposed to provide soft loan at this stage for product development and commercialization at an interest rate of 2%. The role of public R&D institution at this stage too is critical. The partner in the public institution at this stage will get the R&D support as grant.

- Small and medium knowledge-based industries in biotech sector will be encouraged to avail of equity support from the SME Growth Fund of Small Industries Development Bank of India (SIDBI).

(iv) *Code of Best Practice for Disclosure Guidelines:* Setting up a 'Code for Best Practice for Disclosure Guidelines for the Indian Biotech Industry'. This Code will be a part of the General Listing Requirements and Disclosure Guidelines and will be in conjunction with SEBI's General Listing Rules and Disclosure Guidelines.

BIOTECHNOLOGY PARKS AND INCUBATORS

Establishing biotechnology parks for the growth of the biotechnology industry is essential either through public-private alliance or public/private sponsorships. With its large human resource in molecular biology, microbiology, biochemical engineering, synthetic organic chemistry, chemical engineering and allied branches of engineering and strong institutional base at the universities, CSIR, ICMR and ICAR, India is well placed to support a number of biotech parks.

Biotechnology Parks can provide a viable mechanism for licensing new technologies to upcoming biotech companies to start new ventures and to achieve early stage value enhancement of the technology with minimum financial inputs. These biotech parks facilitate the lab to land transfer of the technologies by serving as an impetus for entrepreneurship through partnership among innovators from universities, R&D institutions and industry. Basic minimum components for parks should include research laboratories for product development, multi-purpose pilot facility for manufacturing and process development, quality control and validation of technologies, common effluent treatment plant, a GLP Animal House, a recognized human resource training centre, administrative support centre etc. The biotech parks should be located so as to be easily accessible for all the stakeholders, tenants, academia with connecting roads, water and power supply and should also attract less administrative clearances from the government.

Strategic Actions:

i. The Department of Biotechnology will promote and support at least 10 biotech parks by 2010. Each park will necessarily meet the qualifying criteria related to the characteristics of the location, a viable business plan, management strategy and a clear definition of the partners and their roles.

ii. The Department of Biotechnology will support creation of incubators in biotech parks promoted by a private industry or through public-private partnership in the form of grant upto 30% of the total cost or upto 49% in the form of equity.

iii. It is proposed that a central body Biotechnology Parks Society of India (BPSI) be set up for the promotion of biotechnology parks in the country on the same lines of the Software technology Parks of India (STPI). The BPSI should be run by professionals having experience in the areas of biotechnology, knowledge in Acts and Rules relevant to biotechnology and management skills. The existing parks can become members of these new biotech parks. The BPSI would be responsible for evaluating the project proposals and advising the Department of Biotechnology on the funding pattern; facilitating industries in obtaining industrial, environmental and other relevant approvals from the central government; making recommendation regarding fiscal incentives to be granted to the biotechnology parks; providing guidance to the venture capital institutions on investment in biotech parks; providing accreditation to the parks etc.

iv. Concessions to biotech companies located in biotech parks. Biotech companies located at biotech parks are eligible for benefits as per the recent changes in the Foreign Trade Policy:

- o Duty free import of equipment, instruments and consumables.
- o Tax holiday under Section 10A/10B of the Income Tax Act
- o A scheme will be put in place for operationalising of the incentives to biotech units located in biotech parks. As a part of this scheme biotech company located in biotech parks to be allowed a five-year time frame to meet the export obligation norms under the SEZ scheme. This measure helps to address the long and unpredictable gestational time lines that are inherent to biotech product development.

REGULATORY MECHANISMS

It is important that biotechnology is used for the social benefit of India and for economic development. To fulfill this vision, it has to be ensured that research and application in biotechnology is guided by a process of decision-making that safeguards both human health and the environment with adherence to the highest ethical standards. There is consensus that existing legislation, backed by science based assessment procedures clearly articulates rules and regulations that can efficiently fulfill this vision.

Choices are required to be made that reflect an adequate balance between benefit, safety, access and the interest of consumers and farmers. It is also important that biotechnology products that are required for social and economic good are produced speedily and at the lowest cost. A scientific, rigorous, transparent, efficient, predictable, and consistent regulatory mechanism for biosafety evaluation and release system/protocol is an essential for achieving these multiple goals.

Strategic Actions:

i. The recommendation of the Swaminathan Committee on regulation of agri-biotech products and of the Mashelkar committee on recombinant pharma products will be implemented in 2005

ii. It is recommended that an event that has already undergone extensive biosafety tests should not be treated as a new event if it is in a changed background containing the tested and biosafety evaluated "event". Where adequate evidence is available that the recurrent parent genetic background of a notified/registered genotype is nearly restored (through field data/molecular data), only the agronomic performance and the level and stability of the transgene expression may be analyzed by two-year trial data by the ICAR. Even in case of a structurally altered transgene with no significant modifications in protein conformation, the toxicity and allergenicity tests need not be carried out provided the predicted antigenic epitope remains the same and the level of expression of the transgene is within the defined limits. For the released event, Department is of the view that there is no need of large-scale trials under the Genetic Engineering Approval Committee, as the biosafety aspects have been already addressed adequately before releasing the "event". Only ICAR trials may address the agronomic evaluation of the crop.

iii. An inter-ministerial group chaired by a reputed scientist will be established in 2005 to address anomalies and issues that arise in regulation from time to time. It is proposed that the administrative support to this committee be through the Department of Biotechnology. The mandate of the committee should be to vet any changes in policies, procedures, protocols by departments dealing with regulation in biotech products and processes; resolve issues emanating from the overlapping/conflicting rules in various acts related to regulation of biotechnology activities in R&D, import, export, releases etc. and to review guidelines, protocols, standard operating procedures and ensure their dissemination to all stakeholders from time to time

iv. A competent single National Biotechnology Regulatory Authority be established with separate divisions for agriculture products/transgenic crops, pharmaceuticals/drugs and industrial products; and transgenic food/feed and transgenic animal/aqua culture. The authority is to be governed by an independent administrative structure with common chairman. The inter-ministerial group will evolve suitable proposals for consideration of the government.

v. A centre for in-service training of all professionals, irrespective of their location, engaged in the regulatory process to be established by the Department of Biotechnology in close collaboration with other concerned departments and institutions.

vi. All existing guidelines to be updated and made consistent with the recommendations of the Swaminathan and Mashelkar committees in 2005. New guidelines on transgenic research and product/process development in animal, aqua culture, food, phyto-pharma and environmental application to be put in place in 2005 by the concerned ministries/departments

vii. As an interim measure, a special regulatory cell will be created by the DBT to build capacity in the country for scientific risk assessment, monitoring and management, to foster international linkages, support biosafety research; to obtain and review feedback from different stakeholders and provide support to industry and R&D institutions. This cell will only have a promotional and catalytic role

viii. Measures will be taken to build professionalism and competence in all agencies involved with regulation of biotechnology products

ix. Research is support of regulation, to safeguard health and environment shall be supported by the concerned funding agencies to generate knowledge that will guide regulations and bioethics policy.

x. Concerned ministries will make a vigorous effort to promote acceptance of the Indian regulatory decisions by other trading countries.

PUBLIC COMMUNICATION AND PARTICIPATION

Biotechnology today has become as important as traditional plant and animal breeding have been in the past. At the same time, it raises a number of difficult economic, social, ethical, environmental and political issues that constitute major challenges for the human society. The reception of biotech products by the public has been rather mixed. In general biopharmaceutical products seem to be better accepted than transgenic crops. Clearly it is no longer possible to assume automatic public acceptance of new products and processes that promise public and commercial benefits. Public perception and opinion have a significant influence on the direction and funding of biotechnology research. Hence there is a need to work actively and transparently to inform and engage the civic society in decision-making, and to maintain a relationship of trust and confidence. The government and the industry must actively promote access to information on the benefits and risks in a balanced manner. To achieve this goal, several enabling factors already exist: a sound biosafety regulatory system; well respected appellate and judicial system for redressal of grievances; cadre of willing and able scientists for effective and accurate communication of information; a large body of extension personnel in agriculture, fisheries, veterinary and human health sectors; large NGO network spread across the country; and an effective and independent mass media.

However, several challenges to success need to be recognized while framing the strategies: diverse levels of education and literacy across the country; low understanding of biotechnology among the public; lack of simple communication material; varying quality of science reporting; inadequate inter agency coordination; insufficient dialogue between scientists, industry, policy makers, regulators, consumer for a civil society organizations and the mass media; and lack of sufficiently proactive administrative machinery. There is a need to build public awareness about opportunities and challenges presented by biotechnology development and to inspire public trust and confidence on the safety, efficacy as well as social and ethical acceptability of products among consumers and civil society through the dissemination of accurate information in a coherent, balanced well articulated, user-friendly and transparent manner. Several focused and well-directed measures are needed to achieve public trust and confidence in biotechnology.

Strategic Actions:

(i) Create a cadre of resource persons to reach the stakeholders

- Creation of a cadre of resource persons to provide credible information based on scientific data
- Training media personnel through Institutes of Mass Communication, colleges of journalism and others
- Capacity building among extension personnel in agricultural, fisheries, veterinary and medical sectors
- Involvement of Panchayati Raj institutions in the process of analysis and understanding the risks and benefits associated with GMOs as they will be playing an important role in the local level management of bio-diversity, access to benefit sharing etc.
- Awareness generation among undergraduate and post-graduate students in universities, colleges etc on issues related to biosafety.
- Promoting a genetic literacy movement within government and public schools through 50 genome clubs nature clubs each year.

(ii) Creating a media resource network
- To facilitate access to information

(iii) Empowering policy makers
- Regular training programs for policy makers

(iv) Empowering the judiciary
- Setting up a training school for the judiciary under the aegis of Centre for DNA Fingerprinting and Diagnostics, Hyderabad
- Training through the National Law Schools and other similar institutions

(v) Institutional mechanisms for strengthening public trust
- Establishment of a dedicated training centre for biosafety, food and nutrition safety and standards as per codex alimentarius committee.
- Creation of a 'National Biotechnology Awareness Fund' for providing support for the education and preparation of educational resource material for various sections of stakeholders in different regional languages of the country.

4

AGRICULTURE AND FOOD BIOTECHNOLOGY

INTRODUCTION

Biotechnology is necessary to maintain our agriculture competitive and remunerative and to achieve nutrition security in the face of major challenges such as declining per capita availability of arable land; lower productivity of crops, livestock and fisheries, heavy production losses due to biotic (insects pests, weeds) and abiotic (salinity, drought, alkalinity) stresses; heavy post-harvest crop damage and declining availability of water as an agricultural input. Investment in agricultural related biotechnology has resulted in significantly enhanced R&D capability and institutional building over the years. However, progress has been rather slow in converting the research leads into usable products.

Uncertainties regarding IPR management and regulatory requirements, poor understanding of risk assessment and lack of effective management and commercialization strategies have been significant impediments. India owns very few genes of applied value. The majority of the genes under use-about 40- are currently held by MNCs and have been received under material transfer agreements for R&D purpose without clarity on the potential for commercialization.

The spectrum of biotechnology application in agriculture is very wide and includes generation of improved crops, animals, plants of agro forestry importance; microbes; use of molecular markers to tag genes of interest; accelerating of breeding through marker-assisted selection; fingerprinting of cultivars, land raises, germplasm stocks; DNA based diagnostics for pests/ pathogens of crops, farm animals and fish; assessment and monitoring of bio diversity; in vitro mass multiplication of elite planting material; embryo transfer technology for animal breeding; food and feed biotechnology. Plants and animals are being used for the production of therapeutically or industrially useful products, the emphasis being on improving efficiency and lowering the cost of production. However, emphasis should not be on edible vaccines for which use in real life condition is difficult. Nutrition and balanced diet are emerging to be important health promotional strategies. Biotechnology has a critical role in developing and processing value added products of enhanced nutritive quality and providing tools for ensuring and monitoring food quality and safety.

It has been estimated that if Biofertilizers were used to substitute only 25% of chemical fertilizers on just 50% of India's crops the potential would be 2,35,000 MT. Today about 13,000 MT of Biofertilizers are used-only 0.36% of the total fertilizer use. The projected production target by 2011 is roughly around 50,000 MT. Biopesticides have fared slightly better with 2.5% share of the total pesticide market of 2700 crores and an annual growth rate of 10-15 %. In spite of the obvious advantages, several constraints have limited their wider usage such as products of inconsistent quality, short shelf life, sensitivity to drought, temperature, and agronomic conditions. From a research perspective the spectrum of organisms studied has been rather narrow and testing has been on limited scale and restricted mainly to agronomic parameters. Environmental factors such as survival in the Rhizosphere/phyllosphere and competition of native microbes have not received sufficient attention. Moreover, results on crops are slow to show. Unless there is a policy initiative at the centre and the state to actively promote Biofertilizers and biopesticides at a faster pace, there is unlikely to be a quantum jump in their consumption.

A taskforce headed by Dr MS Swaminthan (2004) under the Ministry of Agriculture has prepared a detailed framework on the application of biotechnology in agriculture. The report rightly lays emphasis on the judicious use of biotechnologies for the economic well being of farm families, food security of the nation, health security of the consumer, protection of the environment, and security of national and international trade in farm commodities.

GUIDING PRINCIPLES

Consistent with the overall vision outlined by MS Swaminathan taskforce, the priorities in agri-biotech would be based on social, economic, ecological, ethical, and gender equity issues. The following guiding principles would apply across the sector:

- A comprehensive and integrated view should be developed of r-DNA and non r-DNA based applications of biotechnology with other technological components required for agriculture as a whole
- Use of conventional biotechnologies (e.g. biofertilizers, biopesticides, bioremediation technologies, molecular assisted grading, plant tissue culture etc.) should continue to be encouraged and supported. A precautionary, yet promotional approach should be adopted in employing transgenic R&D activities based on technological feasibility, socio-economic considerations and promotion of trade.
- Public funding should be avoided to research areas of low priority or those that could reduce employment and impinge the livelihood of rural families.
- Regulatory requirement in compliance with Cartagena Protocol, another international treaty and protocol for biosafety, germplasm exchange and access and the guiding principles of codex alimentarius will be implemented through inter ministerial consultative process
- Transgenic plants should not be commercialized in crops/commodities where our international trade may be affected. However, their use may be allowed for generation of proof of principle, strictly for R&D, their alternate systems are not available or not suitable.
- In a long term perspective basic research for development of low volume, high value secondary and tertiary products through enabling technologies of genomics, proteomics, engineering of metabolic pathways, RNAi, host pathogen interaction and others. Research and support of biosafety regulation would need support.

It is proposed to do away with the large-scale field-testing of the released transgenic events and make it compliant to agronomic test requirements.

Strategic Actions:

(i) Accelerating the pace of product development

- In our quest for better products, strong and sustained support should be given to encourage indigenous discovery of new genes and promoters in both public and privately owned institutions. Nevertheless, wherever there be an urgent need for a product to achieve food or nutritional security, creative commercial and academic international partnerships should be explored in national interest for sourcing important genes and promoters through licensing arrangements on exclusive/non-exclusive basis. The cost effectiveness should be carefully assessed on a case-to-case basis.
- A gene bank should be created and be accessible to private and public sector organizations after payment of an appropriate fee.

(ii) Public-Private Partnership

- There is an urgent need to promote and improve the levels of horizontal integration between public-public and public-private laboratories.
- Institutions that generate knowledge and those that specialize in late stage field trials are currently compartmentalized. While support to public-funded innovation must continue to be strengthened, it is proposed that at least 30% of government-funded programmes must have a commercial partner who will be responsible for directing R&D towards commercialization. Public investment should also be encouraged in small and medium companies, especially for late stage trials of transgenic crops. Partnership between public-funded organizations and industry is crucial in the science-to-product chain.

(iii) Inter-ministerial Agriculture Biotechnology Board

- An inter-ministerial Agriculture Biotechnology Board involving Ministry of Agriculture, ICAR, DBT, MoEF, regulatory authority, expert scientists, industry, and the farming community should be established to continuously assess cross cutting issues such as: duplication of R&D investments; capacity building; promotion of horizontal partnerships between various components in the knowledge-product chain; the most cost-effective manner of overcoming nutrition deficiencies (viz. iron, zinc, iodine, vitamin A); availability, access, release and efficient system for biosafety assessment of GMOs and products thereof; safe use of approved technologies and prevention of unauthorized ones; building public trust and understanding biotechnological application relating to global warming, climate change and sea level rise; global trends in consumer/industry preferences of farm commodities. This will also monitor trade and collect market intelligence with respect to GM crops and products and follow the trend of organic markets and watch international developments to identify niche markets, monitor countries that are rejecting GM foods and feed this intelligence to concerned agencies.

PRIORITIES

Priorities for crops and traits should be set after conducting a need assessment exercise in various farming zones. However, an indicative list has been suggested by MS Swaminathan Task Force (2004).

Crop: Priority target traits in crop plants would be yield increase, pest and disease resistance, abiotic stress tolerance, enhanced quality, and shelf life, engineering male sterility and development of apomixis. Crops of priority should be rice, wheat, maize, sorghum, pigeon pea, chickpea, moong bean, groundnut, mustard, soybean, cotton, sugarcane, potato, tomato, cole crops, banana, papayas and citrus. In priority crops equal emphasis should be given to GM hybrids and new varieties. The varieties in contrast to hybrids, are preferred by small farmers as they can use their own farm saved seeds for at least three or four years. In case of hybrids, research on the introduction of genetic factors for apomixis would be supported so that resource-poor farmers can derive benefits from hybrid vigour without having to buy expensive seeds every cropping season.

Livestock: Priority target traits in livestock would be enhanced fertility and reproductive performance, improved quality, resistance to diseases for reduced drug use, production of therapeutically useful products and quality feed. Livestock of priority would be buffalo, cattle, sheep and goat. Emphasis would be given to animal healthcare, nutrition, development of transgenics and genomics. It is proposed to set up an autonomous institution for animal biotechnology.

Aquaculture and Marine Biotechnology: Application of biotechnology would be crucial in disease resistance, enhanced productivity, fertility and reproductive growth, use of aquatic species as bioreactors for production of industrial products, value added products from sea weeds and other marine taxa and biosensors for pollution monitoring. Species of priority in fisheries would be carps, tiger shrimps and fresh water prawns. It is proposed to set up under the auspices of DBT and autonomous centre for marine biotechnology

Food and Nutrition: R&D would be focused on: development of biotechnology tools for evaluating food safety, development of rapid diagnostic kits for detection of various food borne pathogens; development of analogical methods for detection of genetically modified foods and products derived there from; development of nutraceuticals/health food supplements/functional foods for holistic health; development of pre-cooked, ready-to-eat, nutritionally fortified food for school going children; development of suitable pro-biotics for therapeutic purposes and development of bio food additives. It is proposed to set up (under the auspices of Department of Biotechnology) an autonomous institute for nutritional biology and food biotechnology (2006).

Biofertilizers and Biopesticides: Priorities would include screening of elite strains of micros-organisms and/or productions of super-strains, better understanding of the dynamics of symbiotic nitrogen fixation, process optimization for fermentor-based technologies, improved shelf life, better quality standards, setting up accredited quality control laboratories and standardization of GMP guidelines. Integrated nutrient management system would be further strengthened.

BIORESOURCES

The combined annual global market for the products derived from bioresources is roughly between US$ 500 billion and US$ 800 billion. India is one of the 12 global mega biodiversity centres harbouring approximately 8% of the global biodiversity existing in only 2.4% of the land area. The country is also home to two of the world's 25 hotspots. The varied cultural diversity across the country as well as a very ancient traditional knowledge system associated with the biodiversity represents added assets. Nonetheless, much of this biodiversity is in peril owing, in the main, to anthropogenic causes. Thus, if the goal of converting our bioresources-animal, plant, microbial and marine-into commercially useful products and processes is to be realized, we need to not only conserve the biodiversity and but also utilize it in a sustainable manner.

In this context, absence of a good quantitative information network on bioresources combining remote-sensing data and ground surveys is a major constraint. The situation is even worse for microorganisms. Field-and marine biologists rarely work with molecular scientists and chemists, pharmacologists or other experts, and there is practically no bioprospecting industry. While our traditional knowledge base would be the starting for bioprospecting, ethics and equity should be our guiding principles in benefit sharing.

ANIMAL RESOURCES

India is home to an estimated 86,874 species of animals accounting for 7.25% of global animal diversity. The degree of endemism is high and populations of several animal groups are diminishing due to habitat destruction and poaching. Several species, their products and the services rendered by them are crucial to our economic well-being: pollination services by insects (e.g., honey bees, bumble bees, moths, butterflies, beetles, flies) to our agricultural and forestry crops, honey, silk, lac, musk, skins are just a few examples. Other species (e.g., molluscs, frogs, toads, spiders, termites, and snakes) represent potential reservoirs of useful products such as toxins, venoms, enzymes, therapeutic molecules and other bioactive substances. Prospecting for these and other products should be a priority. Biotechnology should be effectively employed for molecular characterization along with bioscreens in search of useful products. Utilization of selected species as bioreactors for production of complex proteins is another important opportunity.

PLANT RESOURCES

India has a huge treasure of plant resources with over 45,000 known species representing 11% of earth's flora. In terms of flowering plant diversity alone, India ranks tenth in the world. About 33% of flowering plants and 29% of total plants are endemic to the country. Genetic erosion is rampant and conservation is a priority. Prospecting of wild plant resources using molecular approaches and mechanism-based screening should be used to identify novel genes (temperature, drought, salinity tolerant) and gene products (therapeutic compounds, dyes, essential oils, biocontrol agents, gums resins and taxmins).

There are potential ornamentals, including foliage-and flower-bearing plants that could be bulked up to be subsequently cultivated on large scale for domestic and international trade. Bioconversion-both cellular and microbial-should be employed to convert intermediates of secondary metabolism into valued added products. Application of genomics, proteomics and metabolomics in carefully selected plants will be very useful.

Biotechnology can contribute substantially in providing cost-effective therapeutically active biomolecules through target/mechanism-based screens, biotransformation, metabolic engineering and transgenic approaches. Biotechnology should also be utilized to add value to our traditional knowledge especially Ayurveda, Sidha and Unani systems as well as tribal and folk medicine. Medicinal plants are also the prime targets of bioprospecting. Besides, the tools of biotechnology can be used for conservation and characterization of plants.

Fossil fuels are chief contributors to urban air pollution and a major source of greenhouse gases (GHGs)-considered to be the main causes behind the climate change phenomena. In contrast, biofuels are renewable; hence, they can supplement hydrocarbon fuels, assist in their conservation, as well as mitigate their adverse effects on the climate.

Two major biofuels for the transport sector, bioethanol and biodiesel, are fast becoming popular in many countries around the world. While bio-ethanol (called ethanol) is produced from raw materials such as molasses, beet, sugarcane juice, grains and tubers, biodiesel is produced from oil (derived from oil-bearing seeds such as Jatropha curcas, Pongamia pinnata i.e.karanja).

India imports nearly 70% of its annual crude petroleum requirement. The net oil import bill (import minus exports) was Rs 77,058 crore (Rs 770.58 billion) in 2003-04 as against Rs 74,174 crore (Rs 741.74 billion) the previous year. This expenditure on crude purchase impacts the country's foreign exchange reserves in a big way. The petroleum industry now looks very committed to the use of ethanol as fuel.

It is estimated that 75% of the increase in world demand for oil will come from transport. India's transport sector will consume ever-higher amounts of fuel over the coming years. Being one of the largest producers of agro products, including sugarcane, India should take a lead in this worldwide effort at promoting sustainable development.

MICROBIAL RESOURCES

Currently only five percent microbes are culturable but there are others of considerable potential value that need to be characterized by new and novel techniques. The five percent culturable microbes have been a source of valuable products.

India should play a leading role in the study and utilization of microbial resources. Our priorities include: preparation of inventories based on primary and secondary data; exploration of micro flora in the north-eastern region of the country, and extreme habitats (hydrothermal vents, deep sea sediments, highly acidic, alkaline and anaerobic regions, degraded ecosystems etc.) for discovery of novel bioactive molecules; and study, characterization and screening of uncultivable microbes through appropriate molecular approaches.

MARINE RESOURCES

The economic zone of the sea as a source of novel genes and gene products-biopolymers, novel enzymes, new therapeutic leads, and other value-added products such as osmo-tolerant crops-has hardly been explored. Marine organisms also present immense potential as biosensors for pollution monitoring as well as bioreactors for production of novel products. Besides, the study of deep-sea organisms including marine microbes has tremendous implications for human health. Expertise in these diverse areas is scattered across a number of agencies/ institutions. Strategic Actions would be in the following areas.

Strategic Actions:

- There is an acute shortage of expertise in India particularly in taxonomy (the science of the classification of the living and extinct organisms) and microbial ecology. We need to take urgent steps to rectify this.
- Support to capacity building in microbial taxonomy through intensive training programmes at graduate and post-graduate levels
- Promotion of horizontal networking between remote sensing experts, field biologists and computer specialists for inventorisation of bioresources based both on primary and secondary sources of information
- Promotion of closer and effective interaction between biotechnologists, foresters, oceanographers and field biologists.
- Ensure that the use of bioresources be sustainable by regulating the harvesting of medicinal plants
- Formulate a policy to regulate the procurement and sale of medicinal plants in India. Introduce regulatory norms prescribed by DCGI that evaluate the efficacy, safety, and quality of herbal products, which currently are exempt from the scope of any regulation of the DCGI.
- Establish a close working relationship between field scientists, pharmacologists and clinicians so that an all round integration is achieved.

- Public-private partnerships need be promoted for product generation
- Creation of a gene bank for maintaining 'mined' genes
- There is, as on date, only one international depository authority (IDA) in the country at the Microbial Type Culture Collection (MTCC) at IMTECH, Chandigarh; however, for securing our IPR interests, we need to initiate steps to establish a few more centres as IDAs.
- Currently, MTCC does not accept biological materials such as cell lines, cyanobacteria, viruses etc. as it has no expertise or facilities for this purpose. Yet, these are essential for filing patents. IDAs in other countries may refuse to accept such material as they may be potentially hazardous or the shipments may have restrictions. In view of this, the scope of MTCC needs to be expanded by upgrading the existing expertise and infrastructure. Alternately, IDAs should be set up where such expertise and infrastructure are available.
- End products from bioprospecting need to be tested for a variety of parameters before commercial production can begin. There is a need to set up appropriate facilities for such late stage testing of products.
- An autonomous Centre for Marine Biotechnology is proposed to be set up under the auspices of DBT
- An autonomous Institute for Biotechnology for Herbal Medicine under the auspices of DBT is proposed to be established.

ENVIRONMENT

Environmental issues concern everyone. Biotechnology has tremendous potential for application to a wide variety of environmental issues including conservation and characterization of rare or endangered taxa, afforestation and reforestation. It can help in rapid monitoring of environmental pollution, eco-restoration of degraded sites such as mining spoil dumps, treatment of effluents discharged by industries (oil refineries, dyeing and textile units, paper and pulp mills, tanneries, pesticide units etc.), treatment of solid waste, and so on. A number of technologies have already been generated and demonstrated in the country. The real challenge is their adoption by the industry, which has been somewhat uneven.

In general, corporate groups have not been overly enthusiastic in adopting biotechnologies even where they have proven efficacy. The reasons may be several: industry is usually not involved at the planning stage of experiment; enforcement of environmental laws is not always strict or uniform at the ground level and offenders can often escape with impunity; manufacturers frequently change their production schedules based on demand profiles resulting in varied streams of effluents, but microbial consortia specifically designed to one set of effluents may be ineffective in breaking down the changed pollutants. The goal of environmental biotechnology would be to provide cost-effective and clean alternatives for risk assessment and quality monitoring, eco-restoration of degraded habitats, conversion of toxic recalcitrant chemicals into harmless by-products, bioremediation of wastes, value-added products from biomass, control of biological invasion through biotechnological interventions, greener process technologies, and effective ex situ conservation strategies. These can be fulfilled through a deeper understanding-and engineering-of the metabolic pathways for degradation of toxicants, environmental genomics and proteomics, and other molecular techniques.

Strategic Actions: For the diffusion of biotechnologies to be successful the following measures should be put in place:

- Ensuring effective and closer horizontal linkages between research workers and the user corporate groups
- Public-private partnership in research and application of clean technologies
- Strict enforcement of the 'polluter pays' principle. This would require interaction with law enforcement agencies
- Capacity building and training, through workshops, of law enforcement officials, municipal workers, state government functionaries and corporate groups on role and relevance of biotechnology in waste treatment
- Steps to encourage small and medium business companies in producing eco-friendly products, microbial consortia etc. for wider usage

- Building greater awareness for protection of proprietary rights of microbial consortia through appropriate methods (e.g., process patent, trade mark etc.)

Greater inter-agency coordination between DBT, MoEF, ICAR, CSIR, CPCB, user agencies and industry through an inter-ministerial Task Force.

INDUSTRIAL BIOTECHNOLOGY

At present, a third wave of biotechnology-industrial biotechnology-is strongly developing. Industrial biotechnology (also referred to as white biotechnology) uses biological systems for the production of useful chemical entities. This technology is mainly based on biocatalysis and fermentation technology in combination with recent breakthroughs in the forefront of molecular genetics and metabolic engineering. This new technology has developed into a main contributor to the so-called green chemistry, in which renewable resources such as sugars or vegetable oils are converted into a wide variety of chemical substances such as fine and bulk chemicals, pharmaceuticals, bio-colorants, solvents, bio-plastics, vitamins, food additives, bio-pesticides and bio-fuels such as bio-ethanol and bio-diesel. The application of industrial biotechnology offers significant ecological advantages. Agricultural crops are used starting raw materials, instead of using fossil resources such as crude oil and gas. This technology consequently has a beneficial effect on greenhouse gas emissions and at the same time supports the agricultural sector producing these raw materials. Industrial biotechnology frequently shows significant performance benefits compared to conventional chemical technology.

Strategic Actions:

- Focus in industrial biotechnology will be on reducing chemical and toxic load in our effluent streams, developing non-fossil fuels that are eco-friendly and developing green technologies in Industrial processing.
- Encourage public-private partnership to promote investment in this sector.
- Promotion of industrial biotechnology in strategic areas of manufacturing and developing green technologies.

PREVENTIVE AND THERAPEUTIC MEDICAL BIOTECHNOLOGY

A healthy population is essential for economic development. Important contributors to the total disease burden are infections like HIV-AIDS, tuberculosis, malaria, respiratory infections and chronic diseases affecting the heart and blood vessels, neuro-psychiatric disorders, diabetes and cancer. It is important to synchronize the technology and products with the local needs of the health system and to facilitate technology diffusion into health practice. Increasing knowledge about pathogen genomes and subtypes, host responses to infectious challenges, molecular determinants of virulence and protective immunity and novel understanding mechanisms underlying escaped immunity and ways to develop novel immunogens will guide development of vaccines against infectious diseases. Translational research and ability to rapidly evaluate multiple candidates in clinical trials can help accelerate the pace of vaccine development.

New directions in manufacturing and delivery are emerging. Major opportunities to control costs are the more efficient processes for manufacturing of new pharmaceuticals, more efficient systems for production of therapeutic proteins and biomaterials and development of drug delivery systems that release drugs at a target site. A shift from parenteral to oral or transcutaneous administration of drugs and vaccine holds the promise of simplifying delivery in health systems. Medical biotechnology offers a significant possibility for Indian industry to establish a strong pharmacy sector, a growing number of small and medium biotechnology companies, a large network of universities, research institutes, and medical schools and low cost of product evaluation.

The medical biotechnology sector annually contributes over 2/3rd of the biotechnology industry turnover. The Indian vaccine industry has highlighted India's potential by emerging as an important source of low cost vaccine for the entire developing world. Further, economic opportunities through contract research and manufacturing through global partnerships are large if supported by enabling government policies and incentives. The policy goal is to accord high priority to basic and applied research, to strengthen capacity in pre-clinical and clinical product evaluation technologies relevant to all aspects of health and medical care-predictive, preventive, therapeutic and restorative will be supported. Innovation will be supported through new granting mechanisms to support interdisciplinary networks and public private partnerships.

Strategic Actions:

(i) Research emphasis

- Basic and applied research would be supported in molecular and cellular biology, genomics, proteomics, system biology, stem cell biology, RNA interference, host response and new platform technologies.
- Pathogenesis of major diseases and molecular mechanics of disease transmission would be investigated
- Product development will be focused on vaccines, diagnostics, new therapies based on cell and tissue replacement, therapeutic antibodies, herbal medicine, plant based medicine, nucleic acids, therapeutics, drug and vaccine delivery systems, new anti microbial agents
- Research to improve production and manufacturing process and local production of biological reagents for development of diagnostics will be supported.

(ii) Improvements in infrastructure and networks

- A centre for translational research will be established. This new institute will be interdisciplinary and will deal with technology policy for public health, molecular pathogenesis of disease, technology development, scale up, product evaluation and technology diffusion into programmes. Centre will be unique in having a pool of scientists, physicians, engineers, and public health persons working on public health grand challenges. This institute will work through public-private partnerships and be a training centre for product development, IPR and regulation.
- A mission mode programme will be initiated in biomaterial and medical device area as an integrated

effort by the Department of Science & Technology and Department of Biotechnology. The goal is to promote R&D and industrial activity.

- Two centres of molecular medicine will be supported within medical school system closely interacting with basic science institutes.
- A virtual network of stem cell centres will be established, using a city cluster approach to network scientists and clinicians. Two core stem cell research centres will be established together with several network sub-clusters. An umbilical cord stem cell bank will be established.
- Mechanism based screening of herbal drugs known in traditional Indian systems would be carried out so as to get value added therapeutics products quickly
- An inter agency task force of ICMR, Department of Biotechnology, and DST will be established to suggest strategies for strengthening medical school based research. Capacity related to translational biology, clinical trials, molecular epidemiology and product development would be strengthened. Integrated MD-PhD programs will be supported.

(iii) Streamline guidelines and procedures for the approval of recombinant pharmaceutical products. Currently there are multiple regulators, multiple ministries, lack of coordination between these regulators, Over-lapping and duplication of responsibilities of these regulators, lack of a linear progression in the approval process and committees working outside their area of expertise. The Mashelkar Committee (2004) has drawn up a new procedural framework for Biopharmaceuticals, which has streamlined the regulatory process:

- IBSC will monitor all development work (upto 20 Litres) and recommend to RCGM for Animal Toxicity Tests (ATT) & Scale up.
- RCGM will evaluate the recombinant technology & grant permission for scale up-R & D, review and approve for preclinical animal toxicity tests and

evaluate ATT data & recommend to DCGI for Human Clinical Trial (HCT).

- DCGI will permit Human Clinical Trials, review Human Clinical Trial Data, grant permission for Manufacture and Marketing the product and inspect the facility where product is manufactured.
- GEAC will review the manufacturing process to ensure that the LMO (living modified organism) is "inactivated" during the process and send its recommendations to the Drugs Controller General of India within the specified time. The GEAC would confine its approval role to LMOs and Category 3 and 4 microorganisms.

REGENERATIVE AND GENOMIC MEDICINE

The first wave of real healthy life extension therapies seems likely to result from research stem cells and regenerative medicine which helps natural healing processes to work faster, or uses special materials to regrow missing or damaged tissue. Doctors use regenerative medicine to speed up healing, and to help heal injuries that cannot heal on their own.

Regenerative therapies have been demonstrated (in trials or the laboratory) to heal broken bones, bad burns, blindness, deafness, heart damage, nerve damage, Parkinson's and other conditions. Regenerative medicine will result in extended healthy lifespan; we will be able to repair some of the damage caused by aging, organ by organ. The first crop of simple stem cell therapies for regenerative medicine might be only a few years away from widespread availability.

There are major scientific and ethical challenges and safety concerns that must be overcome in taking stem cell based technology for bench to bedside. As it is rapidly evolving field, the existing national (ICMR) guidelines need to be updated and supported by clear articulated procedures. India must consider the potential medical applications of stem cell research. We must reassure end users on the safety and quality by ensuring regulation on stem lines having stable characterizations so that safety risks are predictable. We must reassure suppliers by regulation from lab to market.

Strategic Actions:

- Formulate a comprehensive Human Tissue Act (end 2005) with codes and guidance for regenerative medicine. In the intension, ICMR and DBT will support existing guidelines for stem cell research with clear procedures to be followed by scientists and physician.
- DNA and stem cell banking facilities will be created.
- Lay down clearer laws on animal testing in the country for progress to be made in this sector.
- Emphasize on Intellectual Property Rights, confidentiality and feedback
- Regulation for human tissue engineered products.
- Public awareness to be created in order to allay fears through education programmes, industry conferences and seminars.

DIAGNOSTICS FOR EMERGING MEDICAL PARADIGM

There is potential to generate a new repertoire of tools for screening people for risk of disease, for early detection of infections and chronic diseases and for predicting outcome. In certain circumstances, single tests are requires to detect multiple pathogens or biochemical abnormalities. To be widely useful, diagnostics need to be real time and low cost. Advances in biosensors and gene amplification are in the offing to enable real time medicine. Immuno proteomics has the potential to reveal multiple targets for development of diagnostics for diseases for which existing tools are unsatisfactory. For chronic diseases, a shift from treating disease on an individual basis is visualized by genetic assessment of likelihood of benefit from a therapeutic intervention, the so-called personalized medicine. It is seen that most drugs work in only a proportion of patients, targeting therapy to the right sub-group will not only make therapy more efficacious but also make evaluation of newer products cheaper.

Pharmacogenomics is a rapidly growing segment that provides a wealth of information pertaining to defective or missing genes, which call for differentiated medicine-a new avenue for drug research. This emerging discipline combines both infotech and biotech skills in augmenting high-speed data mining of both genotypic and phenotypic information with a view to evolving new forms of medical diagnostics and therapies. Gene regulation and other bio-algorithms will form the core of a new wave of diagnostics that are now being referred to as 'theranostics'.

India can be positioned as the hub for differentiated medicine as the country offers one of the most affordable development bases for personalized medicines. Personalized therapies will demand extensive clinical data generated from well-differentiated patient populations. India has one of the most desired disease and patient profiles that can enable such studies. Coupled with this is the need for a large number of novel diagnostics based on gene and non-gene based platforms. These are clearly large opportunities for Indian Biotech companies to pursue. Personalized drugs also address the affordability factor for expensive therapies such as those that are involved with cancer. Some important barriers to improving the clinical utility of such knowledge exist. These include the highly complex nature of the problem, little incentive for industry to move to genomic-based approach, and lack of provider education.

Strategic Actions:

- Establish a cell for Diagnostic Biotechnology to encourage and support studies into the clinical application of pharmacogenomics. This cell should be well positioned to overcome barriers in its work to bring pharmacogenomics to the clinical setting.
- Encourage research-involving investigators with both clinical practice and pharmacology/pharmacokinetics expertise while at the same time keeping the overall goal of clinical application/utility in focus.
- Provide incentives for this group of clinician-researchers to bring these scientific advances to the patient bedside
- Support education programs to providers of the importance of this field and its utility.
- Encourage biopharmaceutical companies to include pharmacogenomic data in their drug submissions.

5

NANO BIOTECHNOLOGY AND BIOENGINEERING

INTRODUCTION

Bioengineering covers a wide range of areas such as tissue engineering, biomaterials for therapeutics, biomedical devices and instrumentation, biomedical sensors etc. Tissue engineering, especially of tissues derived from the patient's own cells, offers total acceptance and integration, unlike non-living materials or tissues from other species. Research is focused on developing non-immunogenic materials to serve as scaffolds for regeneration of damaged tissue. Bone and cartilage can be grown today and there is potential for other tissue. Developments in novel biomaterials for micro-particle and nano-particle encapsulated drugs, proteins and other molecules have offered improvement in quality of many therapies with minimal side effects.

Bioengineering offers opportunities for indigenous development of critical implants and extra corporeal devices. Nanoscale structured materials and devices hold a great promise for advanced diagnostics, biosensors, targeted delivery and smart drugs. The application of nanotechnology in bioengineering together with biotechnology offers a great new range of advanced biomaterials with enhanced functionality; and intertwined with tissue engineering, it has the potential to provide true organ replacement technology of the coming decade. While recognizing this potential, it is important to assess not only the efficacy, but also safety of these new interventions regard to human health.

The current market for medicinal devices such as implantables, disposables wound care, dental and orthopaedic materials etc is estimated at around Rs 7000 crores another

Rs 5000 crores for the medical instrumentation sector in the country, with a growth rate of 15% per year. Nearly 80% of this demand is met by imports. Major factors limiting the growth of indigenous medical devices industry are the high cost and non-availability of imported technology, higher risks involved in producing and marketing medical devices, inadequate indigenous technology development and production of biomaterials and device and lack of a regulatory authority for medical devices in the country.

Strategic Actions:

(i) In bioengineering research emphasis will be on:

- Development of tissue engineered skin, cartilage, cornea, acute liver support, large segment bone repair and small diameter artery
- Biomaterials for drug delivery and controlled release
- Regenerative therapy for the failing myocardium through LVAD support, drug therapy and stem cell technology
- Advanced blood compatible surface fir cardiovascular devices
- Advanced burn and wound dressings
- Bioinstrumentation and physiologic monitoring
- Biosensors for detecting and monitoring metabolites and identifying specific genetic materials and for home monitoring of critical parameters like creatinine, cholesterol and triglycerides
- Dental and orthopaedic materials based on polymer-ceramic composites
- Test methods for safety evaluation of tissue engineered and combinational products.

Pre-eminent applications derived from Nano-biotechnology include drug delivery systems and diagnostics. R&D support will be focused on:

- Micro-electro-mechanical systems (MEMS), medical electronics and fibre optics

- Bio-molecular chips for analysis
- Carbon nanotube based biosensors
- DNA nanowire and electrical characterization of DNA

(ii) Establishing effective institutional mechanisms

- An inter agency working group will be formed to develop a common vision and working strategy in this area
- Appropriate regulatory process will be established to hasten introduction of new medical devices through inter-ministerial consultation
- Focused multi-disciplinary research groups shall be formed with clear mandates, targets and adequate funding; these will be monitored regularly for accountability on research output.
- Suitable institution-industry linkage will be built for technology proving and scaling up of products/ medical devices developed at laboratory level

BIO-INFORMATICS AND IT-ENABLED BIOTECHNOLOGY

Bioinformatics has proved to be a powerful tool for advanced research and development in the field of biotechnology. Bioinformatics holds out strong expectations of reducing the cost and time of development of new products such as new drugs and vaccines, plants with specific properties and resistance to pests and diseases, new protein molecules and biological materials and properties. As the full genome sequences, data from micros arrays, proteomics as well as species data at the taxonomic level became available, integration of these databases require sophisticated bioinformatics tools. Organizing these data into suitable databases and developing appropriate software tools for analyzing the same are going to be major challenges. India has the potential to develop such resources at an affordable cost.

Bioinformatics in India can be used effectively for promoting research in biology; prospecting; conservation and management of bioresources; evaluation of products, processes and raw materials, managing complex data required to plan and monitor major national programs; and meeting the growing need of contract services and business outsourcing in pharma and biotechnology sectors. One of the major challenges in optimum exploitation of bioinformatics for solving life science issues is the formulation of appropriate computational biology problems that can be addressed through IT tools. This requires adequate appreciation of the scope and strength of bioinformatics by the biologists and basic understanding of the biological sciences by the information scientists. The solution lies in having adequate leaders with expertise in both life sciences and information technology and strong institutional/program tie-up between specialists from both the fields.

In India, Informatics for life Sciences is an emerging sector- the market size is still quite limited (many verticals each of size USD 20 million-USD 100 million). India has strengths in Chemistry and Computer Science, Software, Health Care and biology.

An extensive bioinformatics network has been established covering more than 60 centres spread all over the country. The network has generated human resources through education and training programs at different levels. Some of them have the potential of emerging as advanced R&D facilities.

To promote R&D and to utilize the business opportunities would require creation of broadband connectivity, high performance computing facilities, virtual reality centres, availability of high quality trained manpower, interactions with bioinformatics centres in different countries and industry academia interactions for joint database and software creation.

Strategic actions:

(i) Human resource development

- A continuous talent pipeline will be ensured by producing 50-100 quality PhDs, 500 M.Sc and 500 advanced diploma holders in bioinformatics every year
- A national testing program will be put in place for accreditation of students at different levels
- The fellowships of PhD students shall be increased

- Industrial training will be introduced for students pursuing advanced diploma course in bioinformatics
- Virtual classrooms will be established in identified institutions. Teaching material in electronic form will be developed and made available at a reasonable cost.
- Industry participation in developing course content and materials will be ensured.

(ii) Infrastructure development

- Super computing facilities with 10 teraflops computing capacity will be created on biogrid to promote protein folding and drug design activities
- Broadband internet connectivity shall be provided for bioinformatics research and manpower development at subsidized rates

(iii) Testing of public domain resources

- Institutional mechanism will be put in place for testing public domain databases and software and making them available to the users from the academia and the industry. After such testing, these databases and algorithms will be graded so that scientists can use them with higher confidence.
- Commercial databases and software will be tested before the industry invests in the products. Such service will help the industry to reduce their costs and use only certified products

(iv) Inter agency coordination

- There are many government departments and agencies, which are supporting bioinformatics activities. These include CSIR, ICMR, ICAR, DST and MIT. An independent inter departmental agency will be established to coordinate these activities among these departments and agencies.
- The agency will be empowered by legislation to provide the direction and oversee the implementation of the coordinated action plan.

(v) Strengthening of DICs and sub DICs

- The CoEs DICs and sub DICs of BTISnet will be strengthened for hardcore research in bioinformatics as well as high-end human resource development.
- Department of Biotechnology will increase the investment in this sector three times over a period of five years

(vi) Bio IT parks and promotion of bioinformatics industries

- Department of Biotechnology in association with the Ministry of IT will set up bio IT parks for the promotion of the bioinformatics industry.
- High-risk projects in bioinformatics will be promoted through special support mechanism including public-private partnerships.

CLINICAL BIOTECHNOLOGY AND RESEARCH SERVICES

Clinical Biotechnology: The cost of launching a new drug into the marked is estimated to cost between $300-500 million of which the cost split between Research and Development is 25%: 75% which would translate to an approximate cost of US$200-400 million for patient clinical studies and trials which form the main components of drug development. The potential of being a key player in this segment is high and remunerative. India has made tremendous progress in clinical biotechnology over the past few years. However, the infrastructure required to identify, document and monitor patients under clinical trials need to be first put in place before India can partake in this activity. There is also an exciting opportunity of conducting longitudinal studies in disease segments for prospecting new biomarkers and novel pharmacogenomic information both yielding high value Intellectual Property.

Research Services: With the global pharmaceuticals companies looking outward to reduce their ballooning research costs, a country like India is in a good position to tap the new business opportunities. Due to the emphasis on outsourcing in the nearly stagnant economies to cut costs and retain competitiveness, India is being considered as a destination for contract research in the pharma sector.

Custom research is a services model that most Indian biotech companies have opted for at their start-up stage in order to earn early revenues with which to fund infrastructure and scientist salaries. These companies harbour ambitions of original R&D once they reach a certain profit level.

Since research service is about delivery of results, the Government has an important role to play in facilitating the industry. India needs to be promoted as a research service destination. The IP environment is confusing since there is no mechanism or standard for contract sharing. The industry needs to collaborate with the academia and a code of conduct for biotech members has to be designed.

Strategic Actions:

- Frame appropriate rules and procedures to support contract research services through stakeholder consultation.
- Harmonise and streamline the regulatory issues for important and export of biological materials.
- Review eligibility of virtual export of R&D services through contract research for fiscal incentives.
- Address the operational deficiencies through stakeholder consultations for conducting clinical trials.
- Develop a Good Clinical Trial Practice Manual taking into account international guidelines and disseminate these widely.
- Promote, train and support clinical trial investigators as a collaborative ICMR, DBT initiative.
- Strengthen clinical trial capacity in medical schools and hospitals and create centres of excellence.
- Address issues and frame guidelines for patent protection including issue of liability.
- Strengthen institutional ethics committees to bring them at par with global benchmark.

INTELLECTUAL PROPERTY AND PATENT LAW

The development of capabilities for the effective management of Intellectual Property (IP) is an important element in securing the benefits of public and private sector research in biotechnology. In this context, filings of patents both in India and aboard are critical to the growth of the Indian biotech Sector. The expenses for filing patents especially outside India are prohibitive and a major barrier to effective Intellectual Property Management within the country.

Whilst expenses incurred with respect to filing of patents in India is eligible for weighted deduction, similar benefit is not provided for expenses incurred with regard to filing patents outside India. As Intellectual Property Right (IPR)* creation is a pre-requisite for exports to the regulated markets, it is recommended that expenditure incurred with regard to filing patents outside India be also eligible for weighted deduction U/S 35 (2AB). This is also imperative in the new WTO-TRIPS regime, which has taken effect on 1st January 2005.

Strategic Actions: Administration of the new intellectual property rights regime should be improved. This will be achieved by:

- Encouraging science graduates to pursue law for better understanding of IPR related issues
- Inclusion of IPR related issues in curriculum of law colleges for facilitating filing of international patents, license negotiation, dispute resolution etc.
- Training scientists and technology transfer professionals in the strategy of intellectual property protection relating to assessment of patentability, prior art examination and technology transfer issues;
- Training patent attorneys on science subject(s) and improving mechanisms for IPR administration through reforms and creation of patent offices, patent codes and ensuring adequate availability of patent attorneys. This will be promoted to an effective inter-ministerial collaboration.

- Setting up of an arbitration council to redress IPR disputes

The setting up of an arbitration council will help in improving the perception and increasing International confidence towards IPR protection in India.

- A Rs 50 crore budget be allocated to substantially improve the current Patent infrastructure and set up additional offices in cities such as Bangalore & Hyderabad.
- The Department of Biotechnology will engage in constant dialogue with the Government of India and WTO-TRIPS to address patentability issues in Biotechnology and their future inclusion in the Patents Bill through ammendments.

The need for an integrated biotech policy with concurrent attention to education, social mobilization and regulation is considered to be an essential pre-requisite for an orderly progress of the biotech sector. Synergy between technology and public policy is essential for us to achieve an effective mobilization of the tools of new biology for adding both years to life and life to years.

The National Biotech Development Strategy has taken into consideration all the areas that will affect the Indian biotechnology industry. The Policy has clearly chalked out direction to strengthen India's academic and industrial biotech research capabilities, work with business, government and academia to move biotechnology from research to commercialization, foster India's industrial development, inform people about the science, applications, benefits and issues of biotechnology, enhance the teaching and workforce training capabilities and establish India as a preeminent international location for biotechnology.

It is imperative that India leverages resources through partnership and build regional innovation systems. The strategy will help develop local talent for a globally competitive workforce. While it recognizes private sector as a crucial player, the strategy also visualizes government to play a major catalyzing role in promoting biotechnology. The development strategy is based on a strong innovation promotion framework in which industry, academia, civil society organizations and regulatory authorities will communicate in a seamless continuum. The perspective for Indian biotechnology would be global while also concentrating on local issues.

HOME GARDEN-TO SAVE HERBS AND TRADITIONAL PLANTS FROM EXTINCTION

A tree lives for about 50 years generates US Dollars 10, 600 worth of oxygen, recycles $ 12,800 worth of fertility and soil erosion control, creates US $ 21, 000 worth of air pollution control, and $ 10, 600 worth of shelter birds and animals. Besides, it provides flowers, fruits and lumber.

"The earth was not given to us by our parents, it was loaned to us by our children."
--Kenyan proverb

Humans have left an impressive mark on the world's lands over the past several centuries. With the dramatic growth in world population, from roughly 1 billion in 1800 to well over 5 billion today, pressures on the land have greatly increased. The need for greater food production has led to a massive increase in cropland. By the early 1990s, almost 40 percent of Earth's land surface had been converted to cropland and permanent pasture. This conversion has occurred largely at the expense of forests and grassland.

The most dramatic changes are occurring in developing countries, where it is estimated that in just three decades--1960 to 1990--fully one fifth of all natural tropical forest cover was lost. Although the forested area seems to have stabilized in developed countries, it is nevertheless only a portion of what was once there. For example, according to a recent estimate, only about 40 percent of Europe's estimated original forest cover remains.

Governments of this region have announced their commitment to saving the tropical forests, while handling out logging concession to their supporters. The rich individuals and local powerful people are destroying the forests and environment. These groups have the power to ride roughshod over forestry departments and to bribe their officials, who themselves are often notorious for their lack of concern for forests or poor. The government officials have to earn "additional income" to keep with the rise of price, whereas the poor people have hardly any opportunity. The poor people see that the outsiders "have stolen trees" and so they have lost faith to protect the forests. The forest departments have developed the same "attitude" as the unpopular police department.

In her speech, the Prime Minister of Bangladesh squarely blamed a section of officials and employees of the forest department for robbing the country of its natural resources that are vital for the survival of very many species of birds, insects, plants and vegetation, and maintaining the thermal equilibrium of the country and the entire region (The Independent, December 11, 2004).

Even as deforestation continues, however, understanding of the value of forests--as regulators of global climate, as repositories of species and potentially valuable new products, as conservators of soil and water resources--is growing rapidly. This increased knowledge has spawned a wide-ranging debate within a variety of international institutions, yet it is still not clear that the world community is ready to forcefully move toward managing forests on a sustainable basis.

If the ecological erosion continues then most of our familiar surroundings and habitual activities will disappear without a trace in geologic time, while there are clearly several realms of human activity-to which we may not give much explicit attention in our everyday lives-that will nonetheless leave indelible and puzzling patterns for future archaeologists to contemplate (Weiskel, 1988).

It was through control of the shattering of wild seeds that humans first domesticated plants. Now control over those very plants threatens to shatter the world's food supply, as loss of genetic diversity sets the stage for widespread hunger.

Large-scale agriculture has come to favour uniformity in food crops. More than 7,000 U.S. apple varieties once grew in American orchards; 6,000 of them are no longer available. Every broccoli variety offered through seed catalogs in 1900 has now disappeared. As the international genetics supply industry absorbs seed companies-with nearly one thousand takeovers since 1970-this trend toward uniformity seems likely to continue; and as third world agriculture is brought in line with international business interests, the gene pools of humanity's most basic foods are threatened.

The consequences are more than culinary. Without the genetic diversity from which farmers traditionally breed for resistance to diseases, crops are more susceptible to the spread of pestilence. Tragedies like the Irish Potato Famine may be thought of today as ancient history; yet the U.S. corn blight of 1970 shows that technologically based agribusiness is a breeding ground for disaster.

When we confront the spread and depth of the diversity of plants that have fed, housed, clothed and cured people all through our existence, we cannot escape that awesome confrontation with time and space. Over an unimaginable number of years plants have evolved and co-evolved with the people who used them; their history and ours, their destiny and ours are intertwined. The open-ended array of soils they have grown in, the hands they have been cared for by, and values they have been fashioned to serve-the diversity of our crop plants is a direct reflection of the diversity of our cultures.

The ancient Romans called it patrimonium, from pater (father). It was used to designate that which was inherited from your father to be transmitted to the next generation, a chain of transmission that could not be interrupted. It was used precisely to distinguish between those goods that could be exchanged for their current monetary value, and those things that had a deeper, inalienable family and community value. Plants definitely fall into this category.

Plants are a fundamental part of the chain of life that keeps this planet going and the diversity within them is the key to their survival. Some of that diversity has evolved through the changing pressures of the environment, but much of it is the result of continuous generations of people tampering with it and passing it on.

We will never be able to measure how much credit goes to 'either side', but there is certainly a part of both. In this sense, genetic diversity is both a natural and cultural heritage that has to be transmitted for the sake of survival. Calling genetic diversity a heritage is not only recognising the role plants play in the chain of life, but also opens up the question as to who is responsible for keeping that chain intact and extending it.

Human communities have always generated, refined and passed on knowledge from generation to generation. Such "traditional" knowledge" is often an important part of their cultural identities. Traditional knowledge has played, and still plays, a vital role in the daily lives of the vast majority of people.

Traditional knowledge is essential to the food security and health of millions of people in the developing world. In mary countries, traditional medicines provide the only affordable treatment available to poor people. In developing countries, up to 80% of the population depend on traditional medicines to help meet their healthcare needs (WHO Fact Sheet No. 271, June 2002).

Medicinal Plants and Local Herbs may be Extinct

Medicinal plants and local herbs may be extinct as the number of students in yunani and ayurvedic colleges remain poor for lack of job opportunities. Practitioners of such medicines said the conservation of such plants and herbs would be difficult without building a group of experts on medicinal plants. "How can medicinal plants be identified without a group of experts," said Hekim Hafiz Azizul Islam, the principal of the Tibbia Habibia Yunani College at Bakshibazar in the capital. He said students are least interested in such alternative systems of medicine as job opportunities remains poor.

"The government created 30 positions of yunani and ayurvedic medical officers in district hospitals about three years ago," a health ministry official said. Two of the 18 yunani and ayurvedic colleges across the country are nationalised. One of the colleges offers bachelor-level course and the remaining colleges offer diploma courses, Hafiz said. All the colleges have a very insignificant number of students. All the 16 private colleges are struggling hard to keep up their existence as they do not often find students. Many of the students drop out as the colleges do not have adequate number of labs and no garden of medicinal plants, the health ministry official said. Students also drop out as the textbooks are mostly written in Urdu,

Persian and Sanskrit, the college principal said (New Age October 18, 2004)

In addition, knowledge of the healing properties of plants has been the source of many modern medicines. The use and continuous development by local farmers of plant varieties and the sharing and diffusion of these varieties and the knowledge associated with them play an essential role in agricultural systems in developing countries.

Only recently, however, has the international community sought to recognise and protect traditional knowledge. In 1981, WIPO and UNESCO adopted a model law on folklore. In 1989 the concept of Farmers' Rights was introduced by the FAO into its International Undertaking on Plant Genetic Resources and in 1992 the Convention on Biological Diversity (CBD) highlighted the need to promote and preserve traditional knowledge. In spite of these efforts which have spanned two decades, final and universally acceptable solutions for the protection and promotion of traditional knowledge have not yet emerged.

Whilst most traditional knowledge and folklore is passed on orally, some of it, such as textile designs and Ayurveda medicinal knowledge, is codified. The groups that hold traditional knowledge are very diverse: individuals, groups or groups of communities may all be custodians. Such communities might be indigenous to the land or descendents of later settlers. The nature of the knowledge is also diverse: it covers, for example, literary, artistic or scientific works, song, dance, medical treatments and practices and agricultural technologies and techniques.

1. Rain Forest Destruction: From the Himalayas to Bangladesh Coastal Plain
2. Bamboo the life blood of the people: Alarm to Ecosystem
3. Plantations (replacing native species) Are Not Forests
4. Ganges Barrage: Ecological Disaster
5. Harmful exotic tree planting still going on
6. The Brahmaputra's Changing River Ecology
7. First Coral Species Listed as Threatened
8. Onslaught on coastal reserve forests

Shal, Shorea Robusta Forests Shrink to 40,590 hectares from 1,20,000 9466 hectare Forestlands Grabbed Illegall

Sal forests in Mymensingh and Tangail districts are disappearing fast because of plunder by thieves and deforestation for pineapple and banana cultivation. There were 1,20,000 hectares of Shal Shorea robusta forests in the central plains and north-eastern regions of the country, according to Abdul Latif Mia, Divisional Forest Officer (DFO), Mymensingh Forest Region. The forests have been reduced to 40,590 hectares now--15870 in Tangail and 7808 in Mymensingh--, according to a government survey done in 1999 and 2000. The DFO also said that 9466 hectares of forestlands have been grabbed illegally in Mymensingh Region.

The government must also take some responsibility because of the foolish decision to hand over forest areas to Eucalyptus, Acasia and Menjiam plantations under a programme entitled "Thana Afforestation and Nursery Development Project" (TANDP). When the Forest Department started the programme, local people including the Garo tribesmen did express resentment at cleaning the Shal forests without taking into consideration the severe environmental consequences that would inevitably follow. In fact any invasive species is a threat to the environment because it can change the entire habitat by crowding out the native species (Editorial, The Bangladesh Observer, December 5, 2004).

The disappearance of forestlands is affecting the environment, bio-diversity and livelihood of tribesmen in Mymensingh, Tangail Jamalpur and Natrakona districts, environmentalists and different NGOs say. A government plan is also responsible for disappearance of Shal forests, sources said.

Although in 1982 the government declared the Bhawal forest a National Park, it did not prevent land grabbers from encroaching on this supposedly protected forest area. Yet the Convention on Biological Diversity imposes an obligation on the government to protect the use of our forests and it is more than time the government takes action against thieves. If, as we believe, the global development agenda set by wealthy countries is also partly responsible for the loss of trees through unwise development programmes, it is important to our welfare that we resist.

The government started plantation of Eucalyptus, Acasia and Menjiam plants in Madhupur Shal forest in Tangail under a programme titled Thana Afforestation and Nursery Development Project (TANDP), funded by the Asian Development Bank (ADB) in 1989. The project was completed in 1995. Later the project area was expended to all four forest ranges in Madhupur and one in Muktagacha forest area in Mymensingh district. Local people including Garo tribesmen had expressed their resentment when the Forest Department had started the programme by cleaning the Shal forests indiscriminately without taking into consideration the severe environmental consequences that would follow, said officials of the Society for Environment and Human Development (SEHD), a local NGO that works on environment. Cleaning of the Shal forests have reduced soil fertility and damaged the environment, they said.

Indigenous people are the worst sufferers because the Shal forests used to provide them with food and shelter. "Shal forests provided us with food and shelter and other requirements for ages. But now the situation has changed totally, keeping us in great difficulties", said Rana Chisim, a student of Madhupur Degree Colleg (A. Islam, Daily star, November 30, 2004).

Forest without Forest Dwellers Development Program by Asian Development Bank (ADB, the World Bank)

Sal Shorea robusta grows well in a well-drained, moist, sandy loam soil. It is a moderate to slow growing species and can attain a height upto 35 m and a girth of about 2 to 2.5 m in about 100 years under favorable conditions. The tree has always been associated with wisdom and immortality. Hindu scripture describes a celestial tree having its roots in heaven and its branches in the underworld that unites and connects beings of all kinds. This is a reversal of our usual experience of trees. However, consider the teaching of the Jewish mystical tradition, Kabbalah. Master Mosheh KHayyim Luzzatto, in the 18th-century classic The Way of God, explains that the higher realms are actually roots that manifest spiritual influence through branches and leaves that permeate the lower realms.Sal leaves and roots increases fertility of soil and protect forest biodiversity. The sal tree is also an object of worship among Buddhists and Hindus in India and the adjoining countries. The legend has it that the famous Lumbini tract where Lord Buddha had sat for meditation and acquired salvation constituted a thick forest of sal trees. It is, therefore, no wonder that some believers treat sal tree as a god.

Sal is the source of an opaline white resin used as incense, as a caulking for boats, and a fuel for lamps. In times of famine, people have been known to grind its fruit for flour, and use its sap to mix with ghee.

Sal Shorea Robusta

The sal wood is considered to be one of the three naturally lasting timbers of the Asian subcontinent,

- *Synonyms:* Sal Tree, Sal, salwa, sakhu, sakher, shal, kandar, sakwa.
- *Family:* Dipterocarpaceae.
- *Range:* Burma in the East, Assam, Bengal, Nepal, Yamuna, Haryana, Shivaliks

Chemical composition: Water = 10.8%. Protein = 8%. Carbohydrate = 62.7%. Oil = 14.8%. Fibre = 1.4%. Ash = 2.3%.

Medicinal Uses: Aromatic oleo-resin gum exuded from stem used to cure cancer, tumour, tubercles, carbuncle, skin infections, syphilis and gonorrhoea; also used as aphrodisiac and stimulant. The resin of shorea robusta is regarded as astringent and detergent and is used in dysentery, and for fumigations, plasters etc.

Other Uses: It is commercial timber species of India. Sal tree also exudes an oleoresin, or ral that is valued as incense in religious ceremonies. It is also used in paint and Varnishes. The presence of resin in the heartwood is responsible for higher calorific value. The Sal seed contains 12-19 percent fat. The fat is used for soap manufacture. After removal of certain ingredients, it is also used as substitute for borea fallow and cocoa butter in the manufacture of chocolates and confectionery. Excellent varnishes can be made from the solution in alcohol. This resin in combination with nitro-cellulose enabled the formulation of rapid-drying lacquers. (Encyclopaedia Britannica, 210A). This project intends to support traditional plants that are threatened by extinction.

While travelling arsenic affected areas in Bangladesh, it was a great surprise, to hear that many arsenic affected women after trying many modern medicine switched to the juice of raw Haldi (turmeric) and experienced astonishingly good results. Use of turmeric is widely known in ancient Ayurveda. There were thousands of useful plants, but current statistics show devastation of vast areas of remaining underdeveloped surface of our world and, with that destruction, the inevitable extinction of thousands of species of plants. In Bengali many of these plants have beautiful poetic names indicating passion for generation to generation. In Bengali traditional plant doctors are called "Kabi Raj", king of the poets!

The Ayurvedic system has described a large number of such medicines based on plants or plant product and the determination of their morphological and pharmacological or pharmacognostical characters can provide a better understanding of their active principles and mode of action.

The basic philosophy of Ayurveda considers that man is an inseparable part of the universe. The human body, mind and spirit continuum is an integral whole and the individual is also linked to the family, society, environment and ultimately the universe. The definition of health is that " It is state of complete psychosomatic equilibrium. It does not mean only absence of diseases but a state in which the mind, senses and spirit are pleasant and active".

Ayurveda is defined as a medical system comprising the wholeness of life's harmony and balance, addressing the dimensions of an individual's physical, emotional, and spiritual balance. Ayurvedic holistic character is, in fact, the one characteristic that allows for analysis of this system as a phenomenon involving from the outset the social, cultural and political forces that influence illness. For example, some of the elements that must be part of any diagnosis are considerations of the familial, social, geographical, and cultural place of the patient, in some cases even complementing a physical examination with a 'land examination'. As Kakar (1982) notes, "the person in Ayurveda, then, is conceived of as simultaneously living in and partaking of different orders of being-physical, psychological, social, and one must add, metaphysical" (Kakar, S (1982): Shamans, Mystics and Doctors: A Psychological Inquiry into India and Its Healing Traditions, Oxford University Press, Delhi.)

According to Samkhya, the philosophical foundation of Ayurveda, creation expresses itself through the five elements: ether or space, air, fire, water, and earth. These elements manifest in the body as the three governing principles or humors called doshas: vata, pitta and kapha. Everyone has all three of these doshas to varying degrees, although one and sometimes two tend to be predominant and the other(s) secondary. In balance, the doshas promote the normal functions of the body and maintain overall health. Out of balance, they create mental, emotional and physical ailments. Vata is the subtle energy associated with movement and is made up of the air and ether. By nature it has dry, light, mobile and cold qualities. When aggravated, it can cause flatulence, constipation, tremors, spasms, asthma, rheumatoid and osteoarthritis, as well many neurological problems. Pitta represents the fire and water elements of the body. It has mainly hot, sharp, and oily qualities. Pitta disorders include hyperacidity, ulcers, skin eruptions, chronic fatigue, Crohn's disease, colitis, gout and numerous inflammatory disorders.

Kapha is made up of earth and water, and is associated with heavy, cold, damp, and static qualities. Out of balance, kapha can cause obesity, high cholesterol, diabetes, edema, asthma, tumors, and a variety of congestive problems.

Aggravation of the doshas can affect the digestion and can create toxins, or ama, to form from poorly digested food. As ama accumulates in the tissues and channels of the body, it slowly but surely affects the flow of prana (vital energy), immunity (ojas) and the cellular metabolism (tejas), eventually resulting in disease.

AYURVEDIC HEALTH-ILLNESS DICHOTOMY

The health-illness dichotomy in Ayurveda refers to two interrelated aspects of a phenomenon: the maintenance of the balance and harmony between environment, body, mind, and soul. Health is always defined as the permanent contest for preserving such a state of balance and wholeness and, ultimately, is its reflection in a high state of consciousness. Illness, on the contrary, beheld primarily as the loss of such balance and harmony, may be caused not only by identifiable diseases in the physical sense of its meaning, but also by mental, emotional or environmental factors. But if these are the general and abstract definitions of the health-illness constructs within Ayurveda, the 'what and how' of these two concepts within this system can be observed in specific issues such as: explanations of the causes of illness, the account of the different stages in which illness is formed, some characteristics of the Ayurvedic physiology and the conceptualisation of the body and, finally, the denotation of prevention as an important mediating concept in Ayurveda.

Like biomedicine, Ayurveda considers viruses and bacteria as causes of illness. But there are intrinsic differences. First, Ayurveda does not see these agents as the only cause of illness. The body and the environment are vast sources of micro-organisms, and it seems simple to say those are the only ones that cause illness. For Ayurveda, all those dimensions that produce health and life, such as soul, mind, senses, and body, could be sources of illness. In Joshi's terms, there are three specific causes of illness: mistakes of the intellect ('pragya aparadha'), misuse of the senses ('asatymya indriyartha-samyog'), and the effect of seasons ('pariman'). Pathogens, then, are only a secondary cause of illness. Second, in contrast to biomedicine's two stages of diagnosis and classification, Ayurvedic discourse explains the manifestation and identification of illness in six stages, called 'shat kriya kal'. Through these six stages it is possible to observe two fully interrelated forces at work: toxicity ('ama') and mobility ('dosha gati'). The first stage is 'sanchaya', which is a period of accumulation characterised by the presence of small imbalances. If these imbalances are ignored or suppressed, illness is invited to progress. The second stage is 'prakopa', which signifies 'aggravation' or 'provocation.' In this stage, if the initial symptoms are not corrected, they will continue growing. The third stage is 'prasara', meaning 'to leave and spread.'

Overflowing of substances and materials are clues for the manifestation of symptoms.

The fourth stage is 'sthana samshraya', or 'taking shelter in a place.' Functional and structural damages are typical of this stage. The fifth stage is 'vyakta', which literally means 'that which can be seen.' This is a stage of clear differentiation in symptoms. In biomedical terms this would be equivalent to diagnosis and classification. The sixth and last stage is 'bheda', or 'differentiation.' Damage and complications are the main characteristics of this stage and, in the worst cases, it leads to death.

Sharp contrast to biomedicine, Ayurveda distinguishes between curable and incurable diseases. As Kakar (1982) notes, "openly listing a number of diseases that are incurable, the 'vaids' [Ayurvedic doctors] do not make the indefensible claim that they can cure all disease" (p 226).

A third contrast between the Ayurvedic body and the anatomical body of western biomedicine is that the Ayurvedic body is a compound of channels with substances flowing through them. In fact, if life is seen as a 'flux', it 'fluxes' through the channels of the body. This is, furthermore, the basic idea of Ayurvedic physiology: to keep all processes flowing through the body's channels. When a channel gets blocked, illness is produced. Illness in one of its conceptualisations, appears as an abnormal process in which the 'flux' is interrupted in a channel. Of course, if a substance stops flowing through its own channel, this creates problems in another channel, contributing to illness.

The fourth and final aspect that allows for observing the relationship between health-illness in Ayurveda comes through the notion of prevention. This is a basic concept for this medical system since it underlines the maintenance of health rather than the treatment of disease. As a naturalistic system that emphasises the rightness of material life processes, Ayurvedic theory insists on the concept that the body evolves through changes and, for this reason, has to be purified and its essences liberated. Indeed, purification and liberation are only possible in a body with open channels through which 'the flux' keeps flowing.

Preventing illness is, therefore, an issue of maintaining the open channels through control of one of the basic material processes of life: eating. It is often stated in Ayurveda that 'who we are is influenced by what we eat.' Thus, food is the key to health and medication is secondary. In Ayurveda, as contrasted with biomedicine, people are not considered passive victims of pathogenic forces but active agents of their quality of life through the choices and interpretations that they make of their bodies and souls. Reducing the consumption of toxins and increasing the use of nourishing substances is, for example, a very simple practice prescribed by Ayurveda doctors for well-being, prevention, and maintenance of health.

From an Ayurvedic perspective, one of the main keys to maintaining optimal health as well as to support the healing process is to help the body eliminate toxins and to reestablish constitutional balance. That agrees with the definition of WHO "Health is a state of complete physical, mental and social well being and not merely the absence of disease or infirmity".

What in the western dichotomy of body-mind is seen as a separation of two aspects that influence each other, in Ayurveda it is seen as a holistic identity that has certain consequences. While biomedicine focuses on the body and illness, it is clear that Ayurveda focuses on the emotional and the person. These body-mind and health-illness dichotomies are vital to understand how a medical system is culturally, historically and politically constructed over time.

Garden can give you Relief

Simple Remedies: For blocked noses, gravel throats and deep coughs: If your nose is all blocked up and you have an excruciating headache, it's because you are unable to breathe properly. Snort a little mustard oil up your nasal passages and while it may sting a little, within minutes you'll be running to blow your nose. Within minutes your nasal passages will be cleared and you'll be able to breathe properly again.

If your throat feels raw and scratchy or you have difficulties swallowing, honey lemon-tea will do the trick.

Add two teaspoons of honey to a cup of raw tea (without sugar) and squeeze in to slivers of lemon. This potion has a two-way effect. The lemon adds flavour and helps to dissolve the honey in the tea that helps to get rid of the scratchy feeling. The raw tea provides the heat necessary to get rid of the swelling. If the problem persists before bedtime, add two teaspoons of honey to warm milk and it will do the same trick while helping you to sleep better.

If you have a deep-set cough that just refuses to come out, stop coughing, it will only make matters worse.

Put a kettle of water on the stove. Add cinnamon, cardamom and some ginger into it. Roll up an old magazine and inhale the steam that comes out of the spout through your nose and mcuth. The heat will travel down and loosen the cough set in your chest. Pretty soon you'll be coughing out big chunks of you know what. In the meanwhile, if your nose is blocked, it will also help to clear it.

If you have allergies that are causing you to itch all over, add neem leaves to your bathing water and you'll find relief. Rubbing a paste of the leaves all over before a shower and then showering helps as well. If the leaves aren't available, you'll find neem oil in the store. Substitute that for the paste to the same effect.

6

BIOTECHNOLOGY AND GLOBALISATION

INTRODUCTION

The story of globalisation of ayurveda is also not the story of opening up of a new world of unlimited opportunities as a result of the rise of the herbal products industry worldwide. A certain kind of opportunities has certainly opened up, but by closing down some other possible openings and by changing the very nature of what was and has come to be recognised as ayurvedic medicine. The change is certainly not for the better. Indeed, there is a case for regarding these changes as downgrading of ayurvedic medicine and reducing it to a more rudimentary form of herbal medicine.

With enormous pressures being exerted by the dominant establishment including the pharmaceuticals industry, alternative medical systems have been confined to marketing alternative products.

The real challenge for ayurveda in the global economy lies in defining the parameters and terms of those parts of its knowledge system that are considered adaptable to the market. However, in the scramble to protect markets and knowledge regimes, it is not yet understood that there is a deeper colonisation being played out in the edging out of alternative world-views inherent in these medical systems (M. Banerjee, EPW, January 3, 2004).

The so-called G21 grouping, in Cancun, which represents more than half the world's population and some two-thirds of its farmers, is united by a common commitment to getting the West to unwind subsidies running at nearly $1 billion a day.

"These are the pressures and blackmail we were going through. They are talking about trade liberalisation and that is their mantra. But then in the areas where they do not have an advantage, like agriculture, they practise protectionism. They have double standards, and the people in those countries need to question their government." (Comment from Uganda, Cancun, 2003). The USA and the rich European countries are trying to create a scenario about which we have read in John Steinbeck's novel The Grapes of Wrath. As this is being written the third draft agreement has also faced opposition from the developing countries. "Developing countries have rejected the EU's anti-development agenda. EU member states such as Britain must now start listening to the emerging opposition of developing countries and address their concerns" (NGOs, Reuters, 2003).

Bangladesh's focus on quota-and duty-free access of LDCs' products to the markets in the developed world appeared to centre around non-agricultural products. Then the issue of movement of semi-skilled workers, which Bangladesh raised so ardently as the coordinator of the LDCs, failed to evoke the desired response (Daily Star, Sept., 2003)

A United Nations report said Friday (30.06.06) that globalisation has failed to close glaring inequalities between rich and poor nations and called for developing countries to be given more space to build up their national economies. The UN's '2006 World Economic and Social Survey' said that inequalities at the global level had grown sharply in recent decades, once fast growing China and India-which between them account for one-third of the world's population-were left out of the equation. It argued that poorer countries need to be given more opportunity to diversify their commodity-based economies to make them less vulnerable to fluctuations in world prices and shocks in international financial markets.

Developing countries should be allowed to implement support measures for nascent export industries, and be granted more special and differential treatment under World Trade Organisation agreements, the report said. Support measures for exports are normally frowned upon in the WTO (Agence France-Presse. Geneva, 2.07.06).

Some Western envoys had expressed skepticism that the G21 would survive long because countries such as Brazil and Argentina, efficient farm goods exporters, appeared to have little in common with India, a protectionist nation of 650 million poor farmers.

Professor Stiglitz, the prime voice against globalisation, has noted in his lectures in Bangladesh and elsewhere in the region that "No agreement is better than a bad agreement." It will be incumbent on the Bangladesh delegation and that of the like-minded LDCs to appreciate and act according to Professor's Stiglitz's views on globalisation.

The late 19th and early 20th centuries also saw a globalisation in economic markets, although not as pervasive and widespread as is the case at the present time. Some claim that globalisation has originated from the dynamics of the phenomenal technological advances, while others, particularly anti-globalisation lobbies, contend that big multinational and transnational companies are the responsible parties for both initiating and carrying the process forward.

The origins of globalisation lie in the political decision by the developed countries, led by the USA with active support from the UK. The purpose, it is suggested, was to develop such an international financial architecture and a trade regime as would mobilise savings from around the world to serve the economic interests of the developed countries, particularly the USA.

Thus, those countries insist that the developing countries open up their markets; but they themselves do not walk the promises made and even the agreements granting preferential market access to imports, particularly to non-agricultural imports, from the developing countries. The level of agricultural subsidy in the USA and the rich EU countries runs at such a high level as US$1 billion a day. On the other hand, they insist that the developing countries do not provide subsidies to agriculture or to any other economic activity.

Moreover, international free movement of labour is not allowed, thereby making the on-going globalisation a partial process, denying the developing countries the one opportunity from which they certainly stand to gain. The Millennium

Developing Goals (MDGs), formulated by the United Nations, focus on issues of deep concern in the developing world. These issues include poverty reduction, promotion of universal primary education, promotion of gender equality and empowerment of women, reduction of child mortality, improvement of maternal health, ensuring environmental sustainability, increasing access to safe drinking water and proper sanitation.

In fact, according to UNDP Human Development Report 2003, 54 countries are now poorer than in 1990. Also, during the same time span, the proportion of people going hungry has increased in 21 countries, and life expectancy at birth has fallen in 34. The free-market globalisation that is being pushed forward by the international dominant powers through such institutions as the World Bank, the IMF, and the WTO and, also, using bilateral mechanisms, virtually discarding the concept of sustainable development for all practical purposes. The issue of environmental sustainability receives a lot of rhetorical attention, but not much in practical terms. Apathy on the part of the rich countries and lack of human and financial capability as well as of far-sight on the part of most of the developing countries are allowing the process of environmental un-sustainability to continue to accentuate.

The USA, the largest contributor to the global emission of greenhouse gases that are responsible for global warming, has withdrawn from the Kyoto Protocol on control of emission of greenhouse gasses, thereby jeopardising the prospect of the Protocol to come into effect.

BIODIVERSITY, TRADE AND DEVELOPMENT LINKAGES

The interlinkages between trade, investment, environment, biodiversity, poverty, rural livelihoods and development are multiple and complex, but very crucial in the unequal but globalized world. The world has seen fundamental and many pervasive changes in the last 50 years. The trends toward globalization has been driven in part by the new technologies and in part by reduced barriers to international trade or trade liberalization and investment flows. The result has been a steady increase in the importance of trade and investment in the global economy while the economy quintupled and the world trade grew by a factor of 14 (IIED and DFID, 2002 and

UNCTAD, 1999). On the other hand, it increased global inequality; the benefits of growth have been very unevenly spread and skewed in favour of the developed northern countries. In many cases trade and investment destructed ecology, biodiversity and livelihood of millions of poor particularly in the least developed and developing southern countries (IISD and UNEP, 2000).

Trade liberalization can also increase exploitation of natural resources and exacerbate the associated negative impactson biodiversity. Despite this, a growing number of developing countries look to trade and investment as a central part of their strategies for development and trade considerations are increasingly shaping their economic and development policy.

Biodiversity also has recreational, cultural, spiritual and aesthetic values. Maintaining biodiversity and access to it, while obviously a planetary public good, is crucial for the poor.

The World Health Organization has estimated that 80% of the world's population depends on traditional medicine derived from local plant varieties for their primary health needs.

Wild plants, in field and forest, make a significant contribution to the diet of many poor communities. In many developing countries, poor communities are able to draw at least half of their food from forest products, and consequently have never faced famine.

The emerging global market forces, technological innovation and commercial interest encourage mono-cropping. High technical input and huge investment backed by commercial interest and chief gains in agriculture and other farm level production have destroyed local knowledge and local resources management practices.

This process seriously affected the natural resources bases and degraded bioresources. This also dislocated millions of marginal and poor people from their traditional occupation and thus affected their livelihood resulting landlessness, poverty, impoverishment in the development countries. Rapid expansion of shrimp farming and huge investment in shrimp sector by the non-resident rich and power elites in the coastal region of Bangladesh has been one of the classical examples of such unsustainable trade and investment.

1. Destruction of sustainable ecosystem for the finest kitchen of the Industrial Countries
2. Poor farmers losing lands to shrimp farm owners.

The process not only dis-benefited poor in terms of their loss of livelihood and reduced access to natural resources and productive assets, but also eroded their capacity and skills in relation to gaining sustainable livelihood, resources management and conservation of biodiversity. Plantation of exotic tree species in the Madhupur forest in Bangladesh dislocating indigenous people could be an example of such bad investment, where few corrupt people and local power elites played a key role in an ADB supported forestry programme.

1. Plantations (replacing native species) Are Not Forests
2. Forest without Forest Dwellers: Development Program by Asian Development Bank (ADB, The World Bank)

Most of the multinationals and global financial institutions such as the World Bank, IMF, ADB have very often supported the commercial production and high technologies. As a result, a small section of people, mainly big merchants, local agents, few government officials, who control the production, processing and exporting of goods and services have been greatly benefited. On the other hand, the common people and the poor are gradually being marginalized and dis-benefited in the unequal and north dominated trade and investment regime. The process not only dis-benefited poor in terms of their loss of livelihood and reduced access to natural resources and productive assets, but also eroded their capacity and skills in relation to gaining sustainable livelihood, resources management and conservation of biodiversity.

The current WTO rules are too deeply influenced by the powerful trading nations, multinationals and liberalization has dis-benefited the developing countries.Many developing countries have criticized Trade Related Intellectual Property Rights (TRIPS), because it willfavour the developed countries and transnational corporations. TRIPS does not provide any guarantee or safeguard to ensure that the poor share in billions of dollars that may be made from the South's biological resources or the application of traditional knowledge; and most importantly, TRIPS reduces farmer's access and control over agricultural resources including seeds which are essential for their food (BCAS, 2004)

The UNCED Agenda 21 suggested to making trade and environment mutually supportive for achieving sustainable development for the global community. The Agenda 21 stresses on poverty eradication, environmental protection and conservation of natural resources and bio-resources.t he great disappointment was the absence of new benchmark, target or timelines in the areas addressed in the action plan (BRIDGES, September 2002).

THE JUDICIAL SYSTEM-SERVES THE INTERESTS OF THE RULING CLASSES

There are many people who believe that the institutionalisation of justice in the form of the judicial system that accompanied the emergence of bourgeois democracy only serves the interests of the ruling classes in this country. For them it is only natur al that the courts will not and cannot give justice to the poor. But for those of us who do believe that the courts-like any other institution in this troubled democracy-are contested arenas for conflicting interests, the only means of ensuring that the courts continue to function as institutions that affirm democracy, is by subjecting them to intense and persistent public scrutiny.

It is now a matter of widespread concern that the judicial system of this country (India) and its judgments are becoming increasingly anti-people: the Bhopal judgment as well as the one on the Narmada are both landmarks of justice denied. The recent judgment on the relocation of small industries and the fate of 25 lakh workers in Delhi becomes incomprehensible when you consider the government data that 67 per cent of all pollution in Delhi comes from vehicles. Yet there is no judgment on the sale, purchase or use of private cars, no real attempt to provide a better public transport system. The rich cannot be touched, but the industries and the workers must go. Does it not need to be asked then, what kind of justice is this, that is so divorced from its real ob jectives and what ends does it seek to meet?The judgment on the Sardar Sarovar Project has to be scrutinised not only because of what it will mean for the millions of people in the Narmada valley who are being uprooted even while it has been made abundantly clear that there is simply no agricultural land available to rehabilitate them. It has to be scrutinised also because it draws larger conclusions about big dams in general (based not on empirical evidence, but on judicial conjecture) and about popular movements and people's access to the system of justice in particular

The majority judgment suggests that people cannot appeal to the courts of law once a project is under way. Since communities are never informed about a project until it begins to be executed, and since this judgment dec rees that they cannot appeal once it begins to be executed, does it mean that affected people can never question a patently bad project? This can only mean that the judicial system-one of the foundations of a democratic system-is unavailable to the common people of this country and their struggles (Front Line, Volume 17-Issue 26, Dec. 23, 2000-Jan. 05, 2001).

Living in a country where Neel Darpan, a celebrated play about the plight of the indigo planters during colonial times, inspired peasant struggles and a resistant subaltern consciousness; where the independence struggle (and many people's movements before and after) were marked by the participation of writers and artists. Understand that the environmental battles of today and tomorrow are not just battles between the Indian elite and the peasantry or workin g class. The battle is between the large mass of common people in this country and global corporatisation. The fight over the control and the use of our lands and rivers is going to be as much in the forests of Madhya Pradesh (where tribal people oppose the World Bank Forestry Project), as on the streets of Andhra Pradesh (where farmers, energy workers and domestic consumers fight the wrecking of the power sector on IMF-World Bank prescriptions), as in the Narmada valley. You will have to decide which side you are on (C. Palit, 2001).

TREE WORSHIP

Tree worship was a part of the religious faith in the prehistoric Indus civilization. The Vedas have praised trees as the sources of herbal medicine needed to fight diseases and for providing wood, as well as food in the pre-agricultural period.

In the Buddhist religion, tree worship had a special place as is evident from the discoveries in places like Sanchi, Barhut, Amaravati, Budhgaya where the mounds or stupas have revealed rich decorative relief work detailing scenes of tree-worship and images of "Brikha" or tree-god.

Trees being nature's major processors of solar energy which is vital for our existence, and yielding flowers, fruit, wood or medicine, have been worshipped by the ancient people of Indian Sub-Continent as a matter of gratitude. Manu believed that they were conscious like humans and felt pleasure and pain. Indian sages and seers eulogized asvattha or peepal (Ficus religiosa), gular (Ficus glomerata), neem (Azadirachta indica), bel (Aegle marmelos), bargad or banyan (Ficus bengalensis), asoka (Sereca indica), amala (Phyllanthus emblica), arjuna (Terminalia arjuna) and many other trees which acquired social and religious sanctity with the passage of time.

It is considered that primitive races of Bengal were tree-worshippers.. A number of trees have become objects of worship. In the popular belief, trees are seen to be favoured by different gods and goddesses who often lived in them. It was this belief that Marmelos tree to Shiva, Tulsi (Basil) to Bishunu, Shal to Durga and Shij to Manesa. In Bikrampur in the District of Dhaka, Hindu women have been worshipping a particular Neem tree by daubing it with vermillion and oil as they think the godess Kali lives in this tree.

Bono bibi or Lady of the Forests-is the presiding female detity of the Sunderbans (Mangrove Forest) cultural zone. She is the gurdian deity of the forest. Both the Hindu and Muslim communities pay their respects before venturing into the forest. Also known as Basuli, Bibima, Bon Durga or Bon Kali, she is potrayed by clay molders either mounted on a tiger or a hen. She is pretty and graceful, ever eager to protect the people of the Snderbans. Our ancestors worshipped the elements: the sun, earth, water, wind, thunder and lightning. The ritual abides; the spirit is gone. We still regard the peepal sacred because the Buddha gained enlightenment meditating under its branches-hence the Latin name ficus religiosa. Its cousin banyan or barh is still worshipped in villages across the country. So is the tulsi (Basil) grown and worshipped in millions of Hindu homes. We worship trees but we do not look after them. We cut down forests every day to cremate our dead. We use wood as fuel to cook and keep ourselves warm. We deprive birds and animals of food and shelter. We must reverse the process, learn to love and cherish our trees.

Berholt Brecht captures man's yearning to establish a close relationship with a tree:

Morning Address to a Tree Named Green

Green, I owe you an apology

I couldn't sleep last night because of the noise of the storm.

When looked out I noticed you swang

Like drunken ape. I remarked on it.

Today the yellow sun is shining in your bare branches

You are shaking off a few tears still, Green.

But now you know your own worth.

You have fought your bitterest fight of your life.

Vultures were taking an interest in you.

And now I know:it's only by your inexorable

Flexibility that you are still upright this morning.

In view of your success it's my opinion today:

It was no mean feat to grow up so tall

In between the tenements, so tall, Green, that

The storm can get at you as it did last night.

Worship of Tree 'Karam Puja'

With a view to starting the festival the 'Thakur' (priest) along with some of the members of their community went to Jonepur, some two kilometres away from Natshal, to cut a branch of Karam or Kadamba tree. There they lighted an earthen lamp (Pradip) and offered worship at the foot of the tree. Then one of them climbed the tree and cut a branch of it. They returned to Natshal, one of the venues of the festival with that branch of the Karam tree and planted it.

The aborigine men and women passed the whole night by singing and dancing surrounding the branch of the 'Karam' with 'Madal' and 'Karatal'. In the morning, they sank the branch in a nearby pond. This was the main ritual the aborigines had long been performing. But there is a story that they believe to be the cause of introduction of Karam Puja.

The aborigines, who live mainly on agriculture, believe that to get proper benefit from agriculture they must worship the branch of Karma (Kadamba) in the name of the 'Karma God'.

The story that they believe is like this: Karma and Dharam were two brothers. Karma worked hard but Dharma did not work. He only worshipped a branch of a tree.

At this, being very angry Karma once threw away the branch which fell on an island across seven seas and thirteen rivers. Karma began to suffer for his neglect of the Kadamba branch and found no more success in agriculture. Karma realised his guilt and after toiling too much took back the branch and started worshipping it. At this he regained his success in agriculture. From that moment 'Dal Puja' or 'Karam festival' came in culture of the aborigines.

Karam festival was actually the festival of the 'Orao' tribe who used to celebrate the festival at their respective areas. Jatiya Adibasi Parishad and Adibashi Sangskritik Parishad jointly started celebrating the festival about eight years ago at Natshal field on the next day of the main ritual. Now it has become a great communion of all the aborigines like Orao, Santal, Munda, Mahato and Raichatri. Sounds of 'dhol' and clapping of the aborigine men, women, old, young and children create a dancing excitement in the blood of all gathered there. But it was closely observed that a section of political personalities have spread their claws to take the minority group under their control. They make the total arrangement of the festival at Natshal field from background although they do not belong to that community (The Independent, September 4, 2004).

Yusuf et. al., (1994) in a recent publication gave a list of 546 medicinal plants that occur in Bangladesh. However, the inventory is not complete, and many plants with medicinal value are yet to be discovered.

The Rangamati Hill District in CHT still harbors a portion of virgin forest. But the procurement of medicinal plants from the wild habitat for professional collectors to make local medicines is unscientific, indiscriminate, and in most cases leads to overexploitation. There is severe depletion of the natural stands, without any provision for the regeneration of species. Some rare species like Ulat chandol (Gloriosa superba), Sarpo gandha (Rawolfia serpentina), and Aswa gandha (Withania somnifera) have become regionally endangered.

Prior to the last two centuries medical practitioners-whether allopaths, homeopaths, naturopaths, herbalists, or shamans-have to know the plants in their areas and how to use them, since many of the drugs were derived from the plants. Plants contain compounds that include a pharmacological reaction in the human body. Plants are very rich in secondary compounds including alkaloids, glycosides, essential oils, and other organic constituents, are responsible for the medical qualities of plants. Alkaloids obtained from vascular plants are among the most important pharmacologically active compounds. They are bitter-tasting organic compounds that are basic (alkaline) in their chemical properties. In Bangladesh and India many bitter spinach varieties that grow in water and land are known as blood cleaning agent and are preferred as appetiser for thr hundreds of years or more.

Man is not only a great inventor and builder, but he has also proved to be the most destructive force ever to appear on the face of the earth. Besides less than ten percent of the population of this planet enjoys all the resources and determines the future course. Statistics show that the devastation of vast areas of remaining undeveloped surface of our world have been destroyed with inevitable extinction of thousands of plant and animal species.

People have recognised the medical value of plants for thousands of years. In the Vedas, which stretch back more than five thousand years mentioned that spices are not only an integral part of culture but also invaluable to cure for every ailment known to man. Even though our earliest ancestors may not have understood how or why certain plants cured specific ailments, they were well aware that plants heal as well nourish. In developing countries today the majority of the population rely on herbal drugs. About two thousands plants are used medically in the Indian-Subcontinent, while three-quarters of the population of China still use herbal medicine. Before synthetic chemicals dominated medicine, as they do to day, roughly 80 per cent of all drugs were derived from plant materials. Chemists eventually developed synthetic version of many drugs, but these man made products would never existed without nature leading the way. In some cases, chemists have not yet learned to duplicate nature. There are stillmany unknown wild tropical plants of our botanical heritage not yet researched or discovered and many potential cures.

JHUM CULTIVATION

The people of North East India and hilly areas of Bangladesh represent a fascinating variety of cultures. Jhum plays an important cultural role in local customs, traditions, and practices, besides offering economic security to farmers. It would be unfortunate if developmental programmes based on misguided opinions about jhum suppress this unique form of agriculture.

Only occupations providing monetary and social benefits perceived by jhumias to outweigh the cultural and security benefits embodied by jhum are likely to gain acceptance. A balanced approach to development that also recognises the merits of jhum is needed. Then, this remarkable form of organic farming may persist into the 21st century.

Jhum as commonly practised by indigenous tribes in North East India. This 'primitive' form of agriculture, according to supporters of "deforestation":

- resulted in serious environmental problems:
- loss of forest cover, erosion of topsoil, desertification, and
- declines in forest productivity.

Others have also decried jhum as an inefficient form of agriculture, an impediment to progress of forestry, and an agent of destruction of biodiversity. Such beliefs have been widespread since British times, and have even resulted in forcible suppression of the practice, oppression and relocation of tribals in Central India and other hill regions. Rapid demographic and social changes have occurred in many tribal societies of North East India. The environmental impacts of jhum cultivation and its role in people's lives have concurrently changed. The conversion of over 80% of the population to Christianity in less than a century (1894-1994) has dislodged the significant role of superstition and mystique in peoples' relationship with their natural environment. A large majority of peoples is tribal and dependent on jhum for its subsistence and livelihood.

Advantage of Jhum Cultivation

- In contrast, studies by ethnologists have tended to view shifting cultivation favourably. It is considered a diversified system, well adapted to local conditions in moist forest and hilly tracts.Others have argued that traditional shifting cultivation may not be as destructive as modern forest exploitation for timber. Clearance of small patches of forest with long fallow periods may even enhance biodiversity in the landscape due to the creation of a variety of habitats. Amidst such contrasting views, there is a clear need for reliable empirical and scientific data on the nature and ecological impact of jhum.
- Jhum cultivation usually involves cutting of second-growth bamboo forests. Since old growth or primary forest is less extensively available and is more difficult to clear, they are cultivated infrequently. The clearing work usually begins in January-February. The slashed vegetation is allowed to dry on the hill slopes for 1-2 months prior to burning in March-April. Crops are sown with the first rains in April in plots that are 1-4 ha in area. Usually, inter-cropping of one or more paddy varieties with 15-20 other crops (vegetables, maize, chillies, gourds, cotton, arum, and mustard) is carried out.

- Studies showed that, far from being primitive and inefficient, jhum is an ingenious system of organic multiple cropping well suited to the heavy rainfall areas of the hill tracts. The economic and energetic efficiency of jhum is higher than alternative forms of agriculture such as terrace and valley cultivation. This is mainly because terrace and valley cultivation needs expensive external input such as fertilisers (which often get leached or lost in the heavy rainfall hill slopes) and pesticides, besides labour for terracing.
- The superiority of jhum cultivation over some forms of sedentary cultivation partly explains the persistence of this form of agriculture in North East India. Other reasons include the economic security provided by jhum and its cultural importance to indigenous tribes. Poor access to markets, capital, and technical knowhow of more commercially rewarding alternatives such as horticulture and cash crop cultivation also hinders the transition to other occupations. Clearly, one cannot do away with jhum assuming it to be a primitive and inefficient system, as attempted in governmental jhum control programmes and new land use policies. Instead, an unbiased understanding of the advantages of jhum is required for proper design and implementation of developmental programmes.
- Erosion of valuable topsoil in the hills due to jhum has been alleged to cause siltation and floods in the plains. Singh has reviewed studies carried out by the Indian Council of Agricultural Research that compared soil erosion from jhum fields with other forms of cultivation on terraces and contour bunds. These studies show that jhum fields cultivated for a single year and abandoned (the most common practice) have less erosive losses of soil than the other forms of settled cultivation.
- Soil erosion is minimised in jhum due to the retaining of rootstocks of bamboo and trees in burned plots, the rapid recovery of weeds and bamboo following abandonment, and the interspersion of forests and fields

on hill slopes. The evidence for siltation of rivers and floods because of soil erosion due to jhum is weak and possibly untenable. Other factors, such as large scale logging for timber extraction, may be responsible to a greater extent for the deforestation and environmental problems in North East India.

CORIANDER CULTIVATION

Poor People's Rich Food

Indian turnip has a round, flattened, perennial, rhizome (cormus), the upper part of which is tunicated like the onion, the lower and larger portion tuberous and fleshy, giving off numerous long, white radicles in a circle, from its upper edge; the under side is covered with a dark, loose, wrinkled epidermis The spathe is ovate, acuminate, convoluted into a tube at the bottom, flattened and bent over at the top like a hood, varying in color internally, being green, dark-purple, black, or variegated with pale-greenish stripes on a dark ground, supported by an erect, round, green, purple, or variegated scape, invested at the base by the petioles and their acute sheaths.

The plant has cne enormous leaf and one spadix annually. It requires hand pollination in Britain. When ripe for pollination, the flowers have a foetid smell to attract carrion flies and midges. This smell disappears once the flower has been pollinated.

The Arum family, Aroidae, which numbers nearly 1,000 members, mostly tropical, and many of them marsh or water plants, is represented in this country by a sole species, Arum maculatum (Linn.), familiarly known as Lords and Ladies, or Cuckoo-pint.

Description: The flowering organs are contained in a sheath-like leaf called a spathe, within which rises a long, fleshy stem, or column called the spadix, bearing closely arranged groups of stalkless, primitive flowers.

The Arum has large tuberous roots, somewhat resembling those of the Potato, oblong in shape, about the size of a pigeon's egg, brownish externally, white within and when fresh, fleshy yielding a milky juice, almost insipid to the taste at first, but soon producing a burning and pricking sensation.The acridity is lost during the process of drying and by application of heat, when the substance of the tuber is left as starch. When baked, the tubers are edible, and from the amount of starch, nutritious. This starch of the root, after repeated washing, makes a kind of arrowroot, formerly much prepared in the Isle of Portland, and sold as an article of food under the name of Portland Sago, or Portland Arrowroot, but now obsolete. For this purpose, it was either roasted or boiled, and then dried and pounded in a mortar, the skin being previously peeled.

This starch, however, in spite of Gerard's remarks, forms the Cyprus Powder of the Parisians, who used it as a cosmetic for the skin, and Dr. Withering says of this cosmetic formed from the tuber starch, that 'it is undoubtedly a good and innocent cosmetic'; and Hogg (Vegetable Kingdom, 1858) reported its use in Italy to remove freckles from the face and hands.

In parts of France, a custom existed of turning to account the mucilaginous juice of the plant as a substitute for soap, the stalks of the plant when in flower being cut and soaked for three weeks in water, which was daily poured off carefully and the residue collected at the bottom of the pan, then dried and used for laundry work.

Constituents: The fresh tuber contains a volatile, acrid principle and starch, albumen, gum, sugar, extractive, lignin and salts of potassium and calcium. Saponin has been separated, also a brownish, oily liquid alkaloid, resembling coniine in its properties, but less active. Arum leaves give off prussic acid when injured, being a product of certain glucosides contained, called cyanophoric glucosides

The dried root was recommended as a diuretic and stimulant, but is no longer employed. The British Domestic Herbal describes a case of alarming dropsy with great constitutional exhaustion treated most successfully with a medicine composed of Arum and Angelica, which cured in about three weeks.

A homoeopathic tincture is prepared from the plant, and its root, which proves curative in diluted doses for a chronic sore throat with swollen mucous membranes and hoarseness, and likewise for a feverish sore throat.

An ointment made by stewing the fresh sliced tuber with lard is stated to be an efficient cure for ringworm, though the fresh sliced tuber applied to the skin produces a blister. The juice of the fresh plant when incorporated with lard has also been applied locally in the treatment of ringworm.

Arum, an esculent edible root, though, trifled literally, used to be considered an occasional vegetable in a Bengali household's food menu. Arum is the only vegetable that survived the flood and rain although farmers in some places lost their produces to the deluge and the downpour. This was the season of cabbage and cauliflower but those had been either washed away or damaged (India-Bangladesh Flood 2004). Arum can survive under water longer than other agricultural products, so it has become the only hope for consumers (T. Maji, October 18, 2004).

The genus Arum (Araceae) is represented by some 20 taxa in Turkey. Having tuberous roots, broadly hastate vigorous leaves, greenish-yellow spathes A. italicum grows in northern Turkey and flowers between April and May and its reddish berry type fruits ripen in July. Containing significant amount of calcium oxalate crystals, oxalic acid and oxalates in addition to volatile and/or easily destroyed irritating substances, Arum taxa are toxic. However, dried or fresh parts thereof are used for food and in folk medicine in Turkey. Tubers and ripe fruits are used in the treatment of rheumatism and hemorroids while the leaves are consumed as a food.

Arum calocasia (Arbi)It is cool, give strength, an appetizer and increases the quantity of milk in mother's breasts It is diuretic causes the formation of excessive wind and phlegm in the body. It increases the quantity of semen, cures plethora and dysentery.

Uses:

- It is mostly used as a vegetable. Although there are many types of arum calocasia, their properties are approximately the same.
- Grinded tender leaves of arum calocasia mixed with powdered cumin seeds cures excessive bile in the body.
- Vegetable of arum calocasia increases the quantity of mother's milk.

Arisæma curvatum, Hook.; Kunth. India: roots eaten.

Although no specific mention has been seen for this species, it belongs to a family where most of the members contain calcium oxalate crystals. This substance is toxic fresh and, if eaten, makes the mouth, tongue and throat feel as if hundreds of small needles are digging in to them. However, calcium oxalate is easily broken down either by thoroughly cooking the plant or by fully drying it and, in either of these states, it is safe to eat the plant. People with a tendency to rheumatism, arthritis, gout, kidney stones and hyperacidity should take especial caution if including this plant in their diet

The root is carminative, restorative, stomachic and tonic. It is dried and used in the treatment of piles and dysentery. The fresh root acts as an acrid stimulant and expectorant, it is much used in India in the treatment of acute rheumatism.

Millions of people in eight districts in Bangladesh are on the brink of starvation. Thousands more face the threat of ill-health and unemployment. Although the government says it has initiated relief measures, the ground realities belie this claim. One of South Asia's most severe droughts, coupled with a 400% increase in the price of essential goods, has left over two million people in north-western Bangladesh on the brink of starvation and forced residents in eight districts to migrate in search of food and employment.

Famine stricken people looking for the Arum roots but this has become rare and expensive (October, 2004).

A near-famine situation the northern districts-Story of Rafique and Others

Arum dracunculus, L. France: starch of root recommended as a famine food for extending bread flour, after removal of acrid element. Ref. PARMENTIER.

Calla palustris, L. France: starch of root recommended as a famine food for extending bread flour, after removal of acrid element. Sweden: unidentified part of plant used in preparation of bread. Bengladesh: greens eaten; roots may be boiled with rice or cooked as curry and may contain chemical component which can irritate mouth and throat. Vernacular name: Water Dragon. Bangladesh: Kachu. Ref. DARLINGTON & AMMAL, DILLINGHAM (1900), PARMENTIER, RAHAMAN.

Colocasia esculenta Schott (syn.Caladium esculentum, Vent.; Colocasia antiquorum, Schott.) (Shortt gives "Calladem esculuntum," the genus and species probably being misspelled). India (Madras Presidency): leaves and leaf-stalks eaten as greens. Kapingamarangi: leaves of the wild taro eaten. Vernacular names-Tamil: Sainmay keeray, Shamay kilangu. Telugu: Chama kura, Chama dumpa;

Pistia stratiotes, L. India: used as a famine food in 1877-1878. Herb is recorded as eaten at other tjmes. China: young leaves eaten cooked. Philippines: used to treat gonorrhea. Plant has a high potash content, and contains stinging crystals. Occurs in great abundance on the surface of stagnant water and slowly-moving streams;

Chemical Composition.:In addition to its acrid principle it contains a large proportion of starch; also, gum, albumen, saccharine matter, calcium and potassium salts, and extractive. When the acrid property is driven off by heat, the root yields a pure, delicate, amylaceous matter, resembling the finest arrowroot, very white and nutritive. That raphides of oxalate of calcium give to the corm its acridity has been asserted by Weber (1991).

Medical Uses:Recommended in flatulence, croup, whooping-cough, stomatitis, asthma, chronic laryngitis, bronchitis, pains in the chest, colic, low stage of typhus, and various affections connected with a cachectic state of the system. Externally it has been used in scrofulous tumors, tinea capitis, and other cutaneous diseases. Its action in the prostration of low fevers with wild delirium is due to its effects upon the cerebral centers. It is reputed useful in cerebro-spinal fever and scarlatina, when delirium is present, when the tongue is swollen, red, and painful, and the buccal membranes inflamed.

Chronic laryngitis, or minister's sore throat, with sudden hoarseness and aphonia, is specifically influenced by arum. It is also useful in ulceration of the larynx and pharynx. It is a good remedy, internally and locally, for aggravated red sore throat. The powdered root may be given in 10-grain doses, increased, if required, to 20 or 30 grains, and repeated every 3 or 4 hours. It may be taken in sweetened mucilage, syrup, or honey. Specific arum, 1/10 to 10 drops. Its specific effects are best obtained by minute doses of the specific arum-1/10 to 1/2 drop doses.

Arum Farming gains Ground in Bangladesh

Arum farming has been gaining popularity in all the seven Upazillas of the district, as this cultivation is bringing profits to cultivators. Previously arum was found in markets of the district and some markets of the other districts. Only a little quantity of arum was supplied to the markets. But now arum is available in the market. Now in all the haats and Bazaars arum is available, as the farmers have started its farming on commercial basis. Talking to Karim Mondol, farmer of village Par of Kendua Union under Sadar Upazilla said he produced, at least, 50 mounds of arum on his five decimal lands. Of the total, he has already sold 15 mounds in the markets at the rate of Taka 280 to Tk 300 per mound. According to Mondol, he will be able to earn Taka 7000 from his products. He had spent about Taka 1000 for purchasing seeds, preparing lands and for other reasons.

Farmers said they usually cultivate seven varieties of arum, which include 'man kachu, pani kachu, gut kachu, kalika kachu, bish kachu, ole kachu and panchamukhi kachu (locally known mukhi)'. Farmers sow arum seeds in the Bengla months of Jaistha and Asar. Cultivation of arum is not very difficult, as the crop needs neither fertilizers nor pesticides. Moreover, the soil and climate of the district is suitable for arum farming (July, 2004).

Home Gardens-Stabilty of Ecosystem

Home or kitchen garden system proposed by Gonzalez (1985) and Allison (1983) is one of the agro-ecosystem that seems to be well adapted ecologically to tropics. Such gardens existed in this sub-continent (India) but due to reduction of land by farmers year to year and introduction of industrial/pharmaceutical products from the cities this valuable heritage is now gradually disappearing.

Tropical homegardens with their large crop species and varietal diversity are regarded as an ideal production system for in situ conservation of plant genetic resources. They are also known to be fields of experimentation and domestication of wild plants. However, garden diversity varies according to ecological and socio-economic factors and/or characteristics of gardens or gardeners. A home garden with an overstory of trees and an understory of a mixture of herbs and small trees permits year-round harvesting of food products, as well as wide range of other products used by the local people, such as firewood, medical plants, spices and ornamentals. Relatively high species diversity provides resource-conserving and ecological sound farming system.

Homegardens as a Special Agroforestry Niche for Women

The cultivation and management of homegardens by women is a widespread phenomenon among settled groups the world over (Buch 1980, Niñez 1985). This is particularly pronounced in Latin America in areas where women do not traditionally till the land, since it provides an agricultural production niche that is seen as an extension of the home. The homegarden is often a way around taboos against tilling the main cropland, and is usually considered an extension of the home as the women's domain. Moreover, by definition such plots are location-specific to the home area, and as such are accessible to women whose mobility may be limited by custom, or by the complex logistics of mixing travel with child care, food processing and food preparation. Homegardens provide an opportunity to intensify labor inputs to increase production, without adding time away from home and within a flexible schedule shaped around other household responsibilities (Chaney and Lewis 1980).

Summary of advantages of commons plantings for women:

- access to land for production
- access to better quality land than they would normally have access to the convenience (efficiency) of local access to concentrated plantings of normally scattered resources "one stop shopping"
- reduction and improved timing of labour inputs, e.g. through rotational labour
- economies of scale through easier fencing, maintenance, protection and marketing of products from concentrated block plantings

- concentrated access to training and assistance
- benefits of the "group learning curve"
- access to credit.

Home Garden benefits Poor People

The people of Domar Upazila; Nilphamari, Bangladesh benefited from afforestation program, besides maintaining ecological balance has created job opportunities for poor people in the upazila. The members of the Samities (committe) look after the saplings and nurture them till they get matured.

The Samity members are given 40 per cent of the money earned from the sale. After cutting down the Bogra trees, saplings of ten percent medicinal tree and 25 percent fruit bearing tree are again planted.

Again, the saplings of flower bearing trees are planted. The process goes on by rotation. At present, trees of different species on both sides of the railway line are increasing the natural beauty. Soil erosion had completely been stopped and the programme had created job opportunities for the rural people. Besides, the government is earning huge revenue (The Independent, November 13, 2004).

Home gardens are considered to be:

- Variable in size and design;
- Respond to local soil type, drainage patterns, cultural preferences, economic standing of the family, family size and age pattern reflecting a multiplicity of both ecological and cultural components;
- Flexible, dynamic, and changing, depending on the needs of the family.

Distribution of plants:

- Low diversity, regularly patterned planting of crops of potential cash values;
- High diversity, irregularly patterned planting of trees, shrubs, herbs etc;
- Low diversity, widely spaced planting of trees, with low grass or bare soil,

- Very high diversity, intercropped planting of ornamental herbs and shrubs,
- Moderate diversity, alternately planted fencerow surrounding the property primarily composed of fruits and fire wood tree species.

Homegardens are more reliable than crops fields for growing trees and vegetables and are important sources of income for the farmers of Bangladesh.

Homegardens are more Reliable than Crops Fields for Growing Trees

Home garden represents the blending of knowledge gained by ecologists studying the dynamics and stability of tropical ecosystems with the knowledge of farmers and agronomists on how to manage the complexities of food producing ecosystems.

Two parallel systems of production forestry exist in Bangladesh: government forests managed by the Forest Department (FD) and privately owned homegardens. Of the country's total land area, about 1.48 million hectares (ha) are designated as government forest land that covers both natural and plantation forests. About 0.72 million ha of land are disignated as unclassified state forests under the control of the Ministry of Land. Homegardens constitute 0.27 ma he and are scattered all over the country. The public forest land, unclassed state forests and homegardens together make up about 17% (2.46 million hectares) of the potential tree growing area of the country the lowest figure of any South Asian country.

From the physical and socio economic points of view, homegardens are more reliable than crops fields for growing trees and vegetables and are important sources of income for the farmers of Bangladesh. It is observed that farmers tend to sell cropland to fight against pauperization, but retain their homegardens unless absolutely unavoidable: Even functionally landless farmers have their own homegardens, where they grow the essential commodies for subsistanc. It is observed that over half of the fruits, vegetables and spices grown in the homegardens are sold to meet family expenses. In Bangladesh farmers spent only 4.8-12.2% of their total labour. In homegarden management, but 26% to 47% of the total family expenses are met from selling homegarden products. During the last 40 years. the relative importance has shifted from the traditional forestry (in the government managed forests) to homegardens in such a way that today about 55% of requirement of timber, fuelwood and bamboo are met from the homegarden sources. Sunderbans, the largest mangrove forest of the world contain many traditional medical plants that can be planted in many wetland areas of Bangladesh:

Species with Possible Development Potential for Homegarden Use

1. Extinction of Crocodiles
2. Migratory and other Birds in Bangladesh in Danger
3. The Wood Sandpiper Tringa glareola Threating to Extinct

A terrain full of verdant trees, plants, herbs and foliage, the Sunderbans is one of the largest intact mangrove forests in the world declared a World Heritage Site by UNESCO in 1997. Thus globally, the Sundarbans is one of great importance. Home to a variety of species, the forest is unique in that many of its plants and animals are not found anywhere else in the world. Over many years land grabbers have harmed this rare ecosystem; and in this way transformation of the Sundarbans from jungles of great biodiversity to wet rice paddy fields occurred causing much damage to its resources.

Over 60 per cent of Sundari trees are dying in the Sundarbans mangrove forest with high salinity prevalent in Khulna and Jessore regions due to severe lack of sweet water flow from upstream points coupled with negative impact of the Farakka Barrage, a leading water expert said.

Prof. Ainun Nishat, Country Director of IUCN, told BSS today that water is being withdrawn in the upstream of Farakka Barrage in the Uttar Pradesh, northern region and Bihar. He said lack of sufficient water not only hampers cultivation but also creates negative impact on fish resources in the rivers. Prof. Nishat said Sundari wood is more valuable than normal wood. Lack of sweet water contents in the Sundarbans mangrove forest kills Sundari trees, he said. He suggested that the country should ensure augmented flow of the Ganges river and this should be diverted to Khulna region.

Sustainable Tourism Centre at Sundarbans

Species	Medical Use
Acanthus sp.	The crushed fruit makes a good blood purifier as well as dressing for boils and snake bite
Ammonia baccifera	Entire plant is used as puragative
Avicennia Sp.	Seeds made into paste to relieve small pox ulceration. Resinous exude used for birth control purposes
Bruguiera eriopelata	Lotion from fruit used for eye. infections Fruit is chewed as betal nut. Young radical used as vegetable
Caesaalpinia nuga	Roots diuretic, used in the treatment of stone.
Cerbera sp.	Fruit when rubbed gives relief from pain of rheumatism. The sap has purgative property. Sap when externally applied against the poisonous effects of fish stings.
Ceriops sp.	Obstetric and haemorrhage cases are treated with an infusion of Ceriops bark.
Ceriops tagal	Roots used as substitute of quinine.
Derris sp.	Seed powder used for bronchitis and whooping cough.
Ipomoea pes-carpas	Leaves used for rheumatism and as an astringent
Hibiscus tiliaceous	Decoction from leaves useful as hair restorers, expectorants, and for treatment of obstinate causes of urine.
Kandelia sp.	Bark forms an ingredient in a mixture given for diabetics.
Lumnitzera sp.	Decoction relieves thirst in infants. Stem decotion used against itches.
Rhizophora sp.	An infusion of the bark of R. muraconta is given for haematuria. Stilt roots some times used as anchor. Root decoction used in blood pressure.

Sonneratia sp.	Fruit made into poultices for sprain and the fermented juice is used to check haemorrhage. Fruits are edible.
T. quallica	Galls and twigs used as astringent and for dysentery.
Tamarix dioica	Bark used as tonic for skin diseases
Thespesia iampus	Roots and fruits used for gonorrhoea and syphilis
Thespesia sp.	An ointment made from seeds kill lice. The leaves furnish a specific active principle for relieving earaches.
Trianthema portulacastrum	Entire plant used for heart disease and anaemia.
Xylocarpus mekongensis	Bark used for dysentery, diarrhoea and as febrifuge

Hydrologists told BSS that the water level fell by three feet at Hardinge Bridge point in December 2004. They said similar fall in the water level was also noticed at Gorai railway bridge point. The hydrologists of Bangladesh Water Development Board (BWDB) said Bangladesh received 145,000 cusec water at the Hardinge Bridge point on December 1, 2004.

They referred to the field reports and said the country received 107,000 cusec water on December 18 which means that the volume of fall stands at 38,000 cusec further. The BSS roving correspondent observed that the water levels in the Padma and the Gorai are falling continuously. Many shoals (chars) have emerged on the waterways of the Padma between Paturia and Daulatdia ghat. Continuous dredging is underway to ensure navigability of the waterway. The fall of water level has caused disruption in the ferry service (The Bangladesh Observer, January 04, 2005).

Tannin

Ceriops decandra, Bruguiera gymnorhiza, Bruguiera parviflora, Rhizophora musconata, Xylocarpus decandra are the valuable trees that produce tannin in the Sunderbans. Studies by the Forest Research Institute, Dhera Dun, India have shown that the spray-dried extract of a blend of ceriops, myrobalans and Acacia nilotica bark contain 65 percent tannin and the blend is suitable for the manufacturing of crust leather. The use of mangrove bark or extract in tanning is locally well-known. and is not commercially used.

The small leather industry of Indian-subcontinent developed Indian vegetable tanned crust over a hundred years ago to preserve the hide in the safest way to suit Indian conditions. The development of leather processing industry was started in Bangladesh in the late 1940s. Until mid 1960s, the leather was dominated by vegetable tanned products for supply to W. Pakistan, Iran and Turkey. Manufacture of wet blue, the chrome tanned semi-processed leather started featuring in 1965. There was a rapid growth of tanning industry in Bangladesh during 1970s and by the end of 70s. In 1999 Bangladesh exported leather and leather goods worth US $ 225 million. Now chrome-tanned processed leather is the shooting star of the export industry at the cost of serious environmental depletion with cancer producing substances.

There is a potential market for vegetable tanned leather products in the industrial countries. Development countries should not destroy their environment for export industry.

Since millions of years from the Himalayas to the dynamic coastal plain of Bengal was rich in panoramic vegetation and wild life. These tropical moist forests were botanically amongst the richest in the Indian sub-continent. The forests are most important as a repository of one of the world's richest of biodiversity

Sunderbans, the largest mangrove forests of the world, was once covered all along the coastal plain of Bangladesh. Had it been maintained, the Bay of Bengal would have turned into one of the largest fish grounds of the world, gained land one third to the present size of Bangladesh, and have protected millions of lives during cyclone storms. The problems of deforestation is mainly political and it can be solved, if poverty focused projects contain the attitude of "by the people and for the people" participation:

1. Our Blue Planet: Extinction of Mangrove Forests
2. Polluted Leather Industry and slums of Bangladesh
3. Poultry feed churned out from tannery waste

4. Plunder of forest resources unabated in Rangamati
5. HILSA Tenualosa ilisha King of Fishes-Going to Extinct?

Neem Azadirachta indica–The Wonder Plant

Sitala Puja-Caitra navaratras: goddess Sitala who is said to reside in the neem tree is propitiated ritually; Pat Gosain festival in Bengal means neem tree worship; neem leaves are eaten on Vaisakha sukla saptami.

"To the best of my knowledge, no plant material with greater activity against abroader spectrum of pest insect species, has yet been found." Dr Martin Jacobson of United States Dept. of Agriculture-Agricultural Research Center in Beltsiville, Madison, USA.

The Neem is being heralded as a tree for solving global problems by the U.S. Department of agricultural. Equivalent products to NeemHit are already registered in U.S.A. (since 1992) and numbers of European countries. The tree has relieved so many different pains, fevers, infections, and other complaints that it has been called "the village pharmacy."

"Azad dhirakat " from the Persian means "Excellent Tree, Noble Tree" referring to the usefulness and the considerable economic importance of the genus. Locally named in Bangladesh as nim, in In dia as nimba, nimuri etc., Nepal as nim, Tibetan as nimpa, traditionally used to make medicine and pesticid1es. Prof. Heinrich Schmutterer, Department of Phytopathology and Entomology working since thirty years on Neem tree and termed, "Neem is the one of the most fascinated trees of the world". In Bangladesh villagers brush their teeth with the Neem branch.

The neem tree (Azadirachta indica) is a tropical evergreen related to mahogany. Native to east India and Burma, it grows in much of southeast Asia and west Africa. A few trees have recently been planted in the Caribbean and several Central American countries

The neem tree has such a variety of medical applications that it is sometimes referred to as the village pharmacy. Now modern research is proving what has been long known by Ayurvedic medicine practitioners: neem is one of the most effective plant medicines in the world. An extremely powerful blood purifying agent and detoxicant, neem is also effective in the treatment of fever, malaria, skin diseases, dental problems, diabetes, tumors, arthritis, and jaundice. It has gained particular attention from scientists seeking a cure for AIDS, not only for its antiviral properties, but also because it boosts the immune system on all levels without destroying beneficial bacteria, unlike synthetic antibiotics.

Azadirachtin being the key molecule, more concentration on Research & developments have been targeted on Azadirachtin only in India and abroad. However now it clearly known that besides Azadirachtin, salannin, gedunin, azadirone, nimbin, nimbidine, nimbicidine, nimbinol, etc.. are also important liminoids which play an excellent synergistic effects on Insects/ Pests.

Tree is considered a good purifier of air, due to its large leaf area. Native of Burma but grown all over Indian Sub-Continnet. Oilcake, obtained from seeds, is used as a fertilizer and manure. Green twigs are used as tooth brushes for cleaning teeth, and as a prophylactic for mouth and teeth complaints. Parts of the plant are used medicinally and the leaves are placed in suit cases to repel insects and to preserve woollens. An extract of leaves is used in tooth pastes and soaps Seeds yield famous margosa oil of disagreeable garlic like flavour. Oil is said to be effective in treatment of leprosy and skin diseases. Also used as a cure for manage in dogs. Leaves in poultice are used for healing of wounds. Ripe fruits are edible. Due to its bitter taste and disagreeable odour, not removed by conventional methods, neem oil has not been utilised on an industrial scale

Oil obtained from neem seed has been found to be suitable for soap making and for hydrogenation. Seed oil is also used as antiseptic and for burning purposes. Stones from fruits are used as beads in rosaries and necklaces. Azadirachtin, a substance isolated from the tree, has been found to have insect repellent and insecticidal properties. Bark yields tannin. Gum exudate from the bark is used in medicines as a stimulant, and for dyeing silk. Bark is useful in fever, nausea, vomitting and skin disease. Bitter principles of neem oil are reported to have been obtained by extraction with alcohol

The main component of the oil is nimbidin which is very bitter. Nimbidin is used for making several pharmaceutical preparations including emulsions, liquors, ointments, medicinal cosmetics such as lotions, shampoos, creams, hair tonics and gargles. Timber is used for agricultural implements and furniture.

In addition to its numerous uses as a healing agent, neem has been receiving much attention for the ecological benefits it provides. For centuries Indians have been mixing neem leaves with stored grains to prevent insect infestation. But neem is not simply a natural alternative to pesticides; increasingly it is being used to reverse desertification and to reduce erosion and deforestation, making it an important weapon in the fight against global warming. Neem's many practical applications make it of enormous interest to anyone concerned about health and ecology. Included are recipes and practical tips that let you enjoy the many benefits of this miraculous plant.

OTHER PRODUCTS

Neemaura is a environmentally friendly natural Neem bitterns,non toxic and safe biodegradable urea coating agent containing Neem Triterpenes inhibit the growth of nitrifying bacteria resulting in delayed transformation of ammoniacal nitrogen into nitrite nitrogen. This ensures slow and continuous availability of nitrogen matching the requirements of crop plants during their life cycle and effectively retards the nitrification of urea.

Neemaura coated urea mineralizes much slower than plain urea at least two to three times under soil conditions by controlling the multiplication process of soil borne bacteria like, Nitrosomanas and Nitrobacter which are responsible for nitrification. Neemaura formulation contains neem bitters and sulphurous compounds, which are mainly responsible for retarding the process of bacterial action and protects urea from leaching, volatilization and also protects crop from insect pest result in higher yields.

Biogrow

Bio-Organic soil enricher made from complete biodegradble organic Neem cake which is manure for green earth. Neem cake is rich in sulphur compounds, in addition to its intrinsic

N.P.K. value, it possess bitter terpenoids such as Epinimbin-A natural nitrification inhibitor. It is a rich source of plant nutrients, growth promoting substances, nitrogen fixers and phosphorous solubilisers which contribute to vigorous growth and high quality yield

Neemgard

The extracted powder from processed neem leaf, which is used in several herbal cosmetic preparations, medicated herbal tea., Health, and hygiene products.Neemgard effectively controls various fungus and pests during storage of seeds. This is also used at nursery level for sowing of seeds and it controls seeds from attack of various fungus and viruses particularly nematodes.

NeemHit Petspray

NeemHit Petspray is a formulation containing neem kernel extract containing azadirachtin for prevention parasite attacks by scabies, eczema, and mange organisms in pets. The protective thin layerAzadirachtin coating is firmly attached to the animal's skin and fur and protect it in a totally natural way, from fleas, lice, ticks, mosquitc, sandfly, species of midge etc and which also act as an antifeedant, deterring and repelling all young fleas and killing flea eggs or larvae. NeemHit petspray totally natural product for the skin, 100% ecologically safe, environmentally friendly with antibecterial, antifungal, antiviral properties, that has value-added benefits, like an active that arrests any inflammation caused by a prior bite, as well as help to improve the condition of the skin and hair.

Chemistry of Ingredients of Neem

Neem plants, as do all other plants, contain several thousands of chemical constituents.Of special interest are the terpenoids are known from different parts of the neem plant. Of its biological constituents the most active and well studied compound is Azadirachtin. However, in most traditional preparations of neem as pesticide or medicine a mixture of neem chemicals are present and provide the active principles. Several different kinds of azadirachtins (A to K) have been isolated, the most abundant of which is Azadirachtin-A.

The neem terpenoids are present in all parts of the plant, in the living tissues. Recently, the site of synthesis and accumulation of the neem chemicals has been identified as secretory cells. Secretory cells are most abundant in the seed kernels. The secretory cells can be seen with iodine solution. Besides the terpenoids, neem also contains more than 20 sulphurous compounds responsible for the characteristic smell of crushed seeds and neem oil.

Toxicity

In toxicological studies carried out in the USA and Germany, different neem product were neither mutagenous nor cancerogenic, and they did not produce any skin irritations or organic alternations to mice and rates even at high concentrations. In another Canadian study, Neem was found to be harmless to Aquatic invertebrates and other non-target species:

> *"To the best of my knowledge, no plant material with greater activity against abroader spectrum of pest insect species, has yet been found."*

The use of Neem in Bangladesh has been dramatically reduced due to destruction of the trees and emergance of chemical industries. Most of Bangladesh was originally forested with coastal mangroves backed by swamp forests and a broad plain of tropical moist deciduous forest (IUCN, 1987). After deforestation the Asian Development Bank (The World Bank) funded aforestation programme selected exotic species from abroad like Eucalytus sp., Dalibergia Sisso, Leucaena leucocephla, Swiietnia macrophulla, and Leucocedha switternia. which grow faster than local natural tress depleting soil and environment. Unfortunately, Neem was not included in the list. About one hundred year ago Neem plant was imported to Africa, where other plants die because of locus/insects attacks but only Neem flourish with wide branches and leaves. I have seen in Sudan only Neem tree survived, whereas other trees perished. Neem tree can easily grow in sandy soils of coastal area of Bangladesh. Leaf, bark, seed and all part of Neem tree contain useful substances that can be taken as tee, oil and prepared medicine remedy dust allergy, fever, skin diseases, rheumatism etc (Roemmming, 1999, Natur). Professor Heinz

Rembold, of famous Max-Planck Institute of Biochemistry, Germany found any side effects of the use of Neem on human and soils do not contain any hazardous substance as a residue after being used as pesticides. In ancient Sanskrit literature (1500 BC) Neem regarded as life saving and disease preventive plant. It belongs to the family Meliaceae.

Neem Seed

Neem has more than 60 valuable compounds. Over 2000 years that Neem based pesticides have been used in India, many complex processes were developed to make them available for specific use (CSE, India, 2000). Azadirachtin A is the most important biopesticide. There are four important components that make Neem seed the king of the bio-pesticide that prevent and kill more than 400 harmful insects, nematode, fungus, bacteria and virus:

- Azadirachtin
- Meliantriol
- Salannin
- Nimbin.

When the insects eat treated plant, the Neem substances drastically influence the important aspect of the life cycle that prevent the insect further reproducing. A rapid interruption occurs in metabolism, growth and hormone system of the insect that the insects can further reproduce. Another effective advantage is that the insects do not develop resistance.

A neem plant gives about 20-30 kg of seeds. The crushed seeds in water can be used as a powerful pesticide. Since the method has to be repeated, a farmer requires two to three trees for his plants. Other pesticidal activity includes of need include: (1) The formation of chitin (exoskeleton) is also inhibited. (2) Mating as well as sexual communication is disrupted.(3) Larvae and adults of insects are repelled. (4) Adults are sterilised. (5) larvae and adults are poisoned.

Neem Oil

Neem oil is an important export item from India Cold pressed oil obtained by traditional grinding method contains most of the useful and biological active ingredients. The oil contains Glyceriden, linol acid, limonoide etc. One kilogram dry seed produces about 100 ml oil.

The oil constitutes antiseptic and many medical properties that can be use as ointment, furniture varnish, shampoo and cosmetic articles. The Neem oil has been successfully used by the scientists in the industrial countries to remove different plant diseases.Neem oil also displays numerous remarkably proven medicinal properties also stimulative and antiseptic effect when used for massage of the body.

It antiseptic properties has been used to particular advantage in the manufacture of special medicated soaps and tooth pastes, in addition to pharmaceutical preparations like emulsions and ointments. Neem oil has been a important ingredient in soap manufacture and neem oil wood treatment for termite free application.

Neem Leaf

That dry Neem leaves in rice, lentils keep away insects, fungicides is a traditional wisdom but this is not used any more. Many dangerous pesticides are used for the conservation of food products (Anwar, 1993). Neem Leaf extract shows an extraordinary reaction against dangerous fungus (Aspergillus flavus). It stops producing Aflatoxin (one of the most dangerous cancer producing substance).

Neem Bark

Neem bark also contains antiseptic properties. Branches are used as tooth brush in the villages. Scientific investigations in Europe have shown that toothpaste prepared from powdered Neem bark has high value for preventive and curative dental treatment. Neem tooth paste is alsobecoming popular in Europe.

Rediscover Traditional Wisdom

Neem is used for curing skin diseases, muscular pain, nail fungus and many uses that still imperfectly known (i.e. birth control). Unfortunately the use of Neem has decreased dramatically in Bangladesh. Firstly, it is regarded as primitive, socially poor and campaign of chemical industry replaced all traditional uses.

Mode of Action

Neem acts as a biopesticide at different levels and ways. This is very important since the farmer is used to the knock out effect of chemical pesticides. Neem does not exhibit this type of effect on pests but affects them in several other ways

Insect Growth Regulation

It is a very interesting property of neem products and unique in nature, since it works on juvenile hormone. The insect larva feeds when it grows, it sheds the old skin and again starts growing. This particular shedding of old skin is the phenomenon of ecdysis or moulting is governed by an enzyme ecdysone. When the neem components, especially Azadirachtin enter into the body of larvae, the activity of ecdysone is suppressed and the larva fails to moult, remains in the larval stage and ultimately dies. If the concentration of Azadirachtin is not sufficient, the larva manages to enter the pupal stage but dies at this stage and if the concentration is still less the adult emerging from the pupa is 100 % malformed, absolutely sterile without any capacity for reproduction.

Feeding Deterrent

The most important property of neem is feeding deterrence. When an insect larva sits on the leaf, the larva is hungry and it wants to feed on the leaf. This particular trigger of feeding is given through the maxillary glands give a trigger, peristalsis in the alimentary canal is speeded up, the larva feels hungry and its starts feeding on the surface of the leaf. When the leaf is treated with neem product, because of the presence of azadirachtin, salanin and melandriol there is an anti-peristalitic wave in the alimentary canal and this produces something similar to vomiting sensation in the insect. Because of this sensation the insect does not feed on the neem treated surface. Its ability to swallow is also blocked.

AGROCHEMICALS IMPORTED POLLUTANTS

Oviposition Deterrent

Another way in which neem reduces pests is not by allowing the female to deposits eggs. This property is known as Oviposition deterrence, and comes in very handy when the seeds in storage are coated with neem Kernel powder and neem oil. The seeds or grains obtained from the market are already infested with some insects. Even these grains could be treated with neem seed kernel extract or neem oil; after this treatment the insects will not feed on them. There will be no further damage to the already damaged grains and at the same time when the female comes to the egg laying period of its life cycle, egg laying is prevented.

Salient Features of Neem

Neem Biopesticide (Emulsifiable Concentrate) is well suited for an " Integrated Pest Management" (IPM) Program because of the following salient features:

- Neem Pesticide is a natural product, absolutely non toxic, 100% biodegradable and environment mentally friend.
- It is suited for mixing with other synthetic pesticide and in fact enhances their action.
- None or lesser quantity of synthetic pesticides need to be used, thereby reducing the environmental load.
- Several synthetic pesticides being single chemical compounds cause easy development of resistant species of pests. Neem consists of several compounds hence development of resistance is impossible.
- Neem does not destroy natural predators and parasites of pests thereby allowing these natural enemies to keep a check on the pest population.
- Neem also has systemic action and seedlings can absorb and accumulate the neem compounds to make the whole plant pest resistant.
- Neem has a broad spectrum of action active on more than 200 spices of pests.
- Neem is harmless to non target and beneficial organisms like pollinators, honey bees, mammals and other vertebrates.

Lucien Biggeault, President, French Support Committee to GK writes:

During my last visit in February (2001), we attended the closing party of a training session on organic home gardening organised in Cox's Bazar by a dozen NGO's under the umbrella of Care. GK had four agriculturists and one agronomist attending this session. I have seen their display and training material and I was surprised to see how good it was on locally made pesticide from Neem leaves and others, protection of certain insects to fight other bad insects, use of home made compost etc....

Research conducted by, among others, United States Department of Agriculture (USDA) has shown that Azadirachtin A offers protection against more than 130 insects, while it is partly active against more than 70 other insects. Since the potential value Neem-based pesticides was recognised, commercial interests have been increased. Suddenly there is spurt of patent applications from scientists and companies-predominantly from industrialised countries-on Neem related products and processes.

1. US Patent on Neem
2. Development-India: Southern States Takes on biopiracy
3. Tribe Accuses Biologists
4. Hoodia cactus: Western drug industry exploits developing countries
5. Kigelia Trees Kigelia Africana

   Kigelia can cure skin cancer called Malignant Melanoma
6. Use of Ferroman Trap instead of harmful pesticide
7. Pelargonium reniforme Effective in Treating Bronchitis, a South African plant: The Rape of the Pelargoniums

Haldi-Turmeric-Curcuma long-The Spice of Happiness and Festival

To most Indians and Bengalis, turmeric or halud (Bengali)/ haldi (Hindi), is a part of growing up, a magic cure-all for the excesses of childhood. A classic "grand mother's remedy', the virulent yellow powder or paste has been applied to scrapes and cuts of generations of children (A. Agarwal and S. Narain,1996). In Bangladesh marriage ceremony is unthinkable without the rubbing of turmeric paste. Halud (Turmeric) is a classical beautiful word in traditional Bengali literature:

"Halud (turmeric) being ground by a turmeric-fair girl

The yellow plate begins to smile passion with colours.."
(by Jasim Uddin))

Haldi (turmeric) is not only the most important spice in Curry, it also known sine hundreds of years to have many medical uses.

Haldi has been used as a wound healing agent can be found in 200-year-old Ayurvedic scripture. These traditional knowledge systems are communicated by word of mouth. Now the western scientists have 'discovered' our traditional bio-resource as a part of profit, as they intend to trade.

The faces of brides are painted with Turmeric as a decorative to make them more glowing. A long and storied history for this ginger-looking, brown-on-the-outside, bright-orange-on-the-inside rhizome. It was listed as a coloring agent in an Assyrian herbal dating back to 600 BCE.

It was used in sacrificial and religious rites in ancient India and China--and is used likewise to this day. In 1280, Marco Polo mentioned in his journals that he saw tumeric growing in the Fukien region of China, "...a vegetable that has all the properties of true saffron, as well the smell as the color, and yet it is not really saffron." Throughout medieval times, it was known in Europe as "Indian saffron" because of its coloring power, which, incidentally, is terrific. It's the yellow that puts the yellow in ballpark mustard--not to mention in curry

- Curcumin, in tumeric, is an anti-inflammatory component that is helpful for arthritis. It not only helps with rheumatoid arthritis but it improves morning stiffness, the ability to walk for long intervals and is beneficial in diminishing joint swelling.
- Tumeric has antibacterial making it ideal for healing wounds. In cases of acne Turmeric can be made into a poultice and applied to the affected area or taken internally

- As an anti-fungal, it can be used for Athlete's Foot by making a paste. When combined with Ginger, it can be used for ringworm
- Liver protection is another of its marvelous qualities. Tumeric stimulates the flow of bile, lessening the possibility of gallstones. If one is exposed to environmental toxins (and who isn't), Turmeric will help break down the harmful substances. Chemotherapy patients and those consuming alcohol benefit from this wonderful herb. It has also been used to clear up diarrhea/dysentery. People suffering from Hepatitis C also rely on Turmeric for its beneficial effects on the liver.
- Anti-cancer activity, as shown in Rutgers University, research demonstrates it is the curcumin in Turmeric that helps prevent tumor development in their animal studies. Similar studies advocate Tumeric for limiting growth of already formed tumors and may have the potential to deter other cancers such as breast, skin, stomach and colon.
- Turmeric has found to be antioxidant, antibacterial, anticoagulant, antifungal, anti-inflammatory and antiviral. While there are very few contraindications, too much of anything is not a good thing.

The use of haldi (turmeric) comprises:

- Increase immune system
- Treating musculo-skeletal disease
- Wound healing
- Dye
- Detect and warn cyanide adulterated food products
- Liquid seasoning compositions
- Metal colour complexes
- Tinted pit and fissure sealant
- Colouring process and composition for food and beverages

- Process for producing water and oil as a colouring agent
- Conservation food (fish, meat, cooked food);
- Improve intestinal haemorrhage, bowel function, stops irritating substances, reduces fatty compounds and perhaps stops cancer etc

Now about 12 patents have been registered in the USA Although turmeric has long been used in India as a traditional medicine for treatment of various sprains and inflammatory conditions (Indian Journal of Medical Research, 1982). A lack of regulations is allowing foreign scientists to claim exclusive ownership of traditional medicine that we have used for centuries.

STEALING SPECIES

The development of effective cures and preventives against life threatening diseases has been predominant concern of medical researches and scientists. These have been always done in the name of public interest or for the sake of humanity. In the past a successful new drug would invariably bring in profits and recognition for those who were behind it, but their motivation was to serve the people and not personal enrichment. This has changed during recent times and the emphasis is now on profits, It is now common to see that many of the new discoveries becoming the private properties of individuals and firms with the help of intellectual property laws that grant patents over them. This is not confined to new drugs and their manufacturing processes, but has extended to therapeutic methods and even disease-causing (pathogenic) organisms. These have brought about a large number of hitherto unknown and unimagined problems, stifled a lot of research, delayed critically important research, increased the prices of drugs and diagnostic kits. The end result is that thousands of people have been denied the opportunity to live a healthy life.

The bitter gourd is found in many Asian countries where it is used both as a vegetable and as a medicine. In Sri Lanka, it is used to treat skin ailments and diabetes. It has been used in China since ancient times to treat infections and tumours. They have been eating the ripe fruit with the seeds to treat these illnesses. The effectiveness of these ancient medicinal plants in treating virus diseases had been subjected to a lot of research by Chinese scientists.

Among them has been Lee-Huang, who subsequently went to the USA and then started patenting the work she was involved in when a China. Lee-Huang herself had admitted in an interview to Bio-world Today (Age old folk remedies resurface as recombinant anti-HIV, anti-tumour therapeutics by Davil N. Leff 23.10.1996) that bitter melon had been widely eaten in China and had been used in China and South-East Asia for centuries as an anti-infection and anti-tumour agent.

Therefore, these patents that cover the MAP-30 protein are another example of biopiracy and the victims are China and the other Asian countries.

Western countries are plundering the Third World's genetic resources (Biodiversity Conservation: The Threat to Ecological Conservation from Commercial Interests, Vandana Shiva, (Third World Network) 1990):

- Wild species of plants and animals (many of which come from the Third World) contributed $340 million per year to the US between 1976 and 1980. The total contribution of wild species to the US economy has been estimated at $66 billion-more than the total international debt of Mexico and the Philippines combined.
- A wild tomato variety taken from Peru in 1962 has contributed $8 million a year to the US tomato-processing industry. None of these profits or benefits have been shared with Peru
- The periwinkle plant from Madagascar is the source of 60 alkaloids which can treat childhood leukemia and Hodgkin's Disease. Drugs derived from this plant bring the US $160 million a year. And another plant, Rauwolfa Serpentina, from India is the base for drugs which sell for up to $260 million a year in the US alone.
- The value of the South's genetic material for the pharmaceutical industry ranges from an estimated $4.7 billion now to $47 billion by the year 2000.

BENGALI COOKING

India Robbed Again

First it was the neem tree, then turmeric, now another Indian medicinal plant is the target of foreign patents. Ashwagandha (Withania somnifera])has been used for thousands of years in the Ayurvedic system as an aphrodisiac, diuretic and for restoring memory loss.

Officials at the Department of Science and Technology (DST) said "one thing which is very obvious is that Ashwagandaha is catching the attention of scientists, and more and more patents are being filed and granted on it by different patent offices around the world. Seven American and four Japanese companies have filed or have been granted patents on Ashwagandha (Source:Diverse Women for Diversity,Norfolk Genetic Information Network PTI 15 May, 2001).

Ashvagandha Withaia Somnifera grows in drier region of India and is also cultivated. A small or middle-sized undersurb, up to 1.5 m high, stern and branches covered with minute star-shaped hairs.

Drug and Properties

The drug consists of the dried roots of the plant. The antibiotic and antibacterial activity of the roots as as wells as leaves have recently been shown experimentally (Jain, 2001):

- Useful in consumption, sexual, general weakness and rheumatism;
- Diuertic i. e. it promotes urination, acts as a narcotic and removes functional obstruction of body;
- The root poweder is applied locally on ulcers and inflammations.

South Africa's Floral Heritage Sold Off

South Africa's National Botanical Institute (NBI) has sold the rights to develop new strains from national flora to US based company, Ball Horticultural. The unnamed government official who blew the whistle on the deal, which was signed two years ago, said "this effectively hands over South Africa's floral heritage to a US company in exchange for a pittance in royalties".

Battle for Basmati Rice

Basmati is a variety of rice from the Punjab provinces of India and Pakistan. The rice is a slender, aromatic long grain variety that originated in this region and is a major export crop for both countries. Annual basmati exports are worth about $300m, and represent the livelihood of thousands of farmers.

The "Battle for Basmati" started in 1997 when US Rice breeding firm RiceTec Inc. was awarded a patent (US5663484) relating to plants and seeds, seeking a monopoly over various rice lines including some having characteristics similar to Basmati lines. Concerned about the potential effect on exports, India requested a re-examination of this patent in 2000. The patentee in response to this request withdrew a number of claims including those covering basmati type lines. Further claims were also withdrawn following concerns raised by the USPTO. The dispute has however moved on from the patent to the misuse of the name "Basmati."

In some countries the term "Basmati" can be applied only to the long grain aromatic rice grown in India and Pakistan. RiceTec also applied for registration of the trademark 'Texmati' in the UK claiming that "Basmati" was a generic term. It was successfully opposed, and the UK has established a code of practice for marketing rice. Saudi Arabia (the world's largest importer of Basmati rice) has similar regulations on the labelling of Basmati rice. The code states that "the belief in consumer, trade and scientific circles [is] that the distinctiveness of authentic Basmati rice can only be obtained from the northern regions of India and Pakistan due to the unique and complex combination of environment, soil, climate, agricultural practices and the genetics of the Basmati varieties."

But in 1998 the US Rice Federation submitted that the term "Basmati" is generic and refers to a type of aromatic rice. In response, a collective of US and Indian civil society organizations filed a petition seeking to prevent US-grown rice from being advertised with the word "Basmati". The US Department of Agriculture and the US Federal Trade Commission rejected it in May 2001. Neither considered the labeling of rice as 'American-grown Basmati' misleading, and deemed "Basmati" a generic term. The problem is not just limited to the US; Australia, Egypt, Thailand and France also grow basmati type rice and may take the lead from the US and officially deem "basmati" a generic term. The name "Basmati" (and the Indian and Pakistani export markets) can be protected by registering it as a Geographical Indication. However, India and Pakistan will have to explain why they did not take action against the gradual adoption of generic status of basmati over the last 20 years. For example, India did not lodge a formal protest when the US Federal Trade Commission formally declared "basmati" generic.

Bitter gourd Karela

The bitter gourd is a common vegetable cultivated extensively all over Indian Subcontinnent. It is 10 to 20 cm. long, tapering at the ends and covered with blunt tubercles. The seeds are white in raw fruits and become red when they are ripe. There are two varieties of this vegetable. The large kind is long, oblong and pale green in color. The other kind is small, little oval and dark green. Both the types are bitter in taste. They turn reddish-orange when ripe.The original home of bitter gourd is not known except that it is a native of the tropics. It is widely grown in India, Indonesia, Sri Lanka, Malaysia, the Philippines, China and the Caribbean.

Natural Benefits and Curative Properties

The bitter gourd has excellent medicinal virtues. It is antidotal, antipyretic tonic, appetizing, stomachic, antibilious and laxative.' The bitter gourd is also used in native medicines of Asia and Africa.

Karela (bitter gourd) Subji-Vegetable Curry

Diabetes: The bitter gourd is specifically used as a folk medicine for diabetes. Recent researches by a team of British doctors have established that it contains a hypoglycaemic or insulin-like principle, designated as 'plant-insulin', which has been found highly beneficial in lowering the blood and urine sugar levels. It should, therefore, be included liberally in the diet of the diabetic. For better results, the diabetic should take the juice of about four or five fruits every morning on an empty stomach. The seeds of bitter gourd can be added to food in the powdered form. Diabetics can also use bitter gourd in the form of decoction by boiling the pieces in water or in the form of dry powder.A majority of diabetics usually suffer from malnutrition as they are usually under-nourished. Bitter gourd being rich in all the essential vitamins and minerals, especially vitamin A, B1, B2, C and Iron, its regular use prevents many complications such as hypertension, eye complications, neuritis and defective metabolism of carbohydrates. It increases body's resistance against infection.

Blood Disorders

Bitter gourd is highly beneficial in the treatment of blood disorders like blood boils, scabies, itching, psoriasis, ring-worm and other fungal diseases. A cupful of fresh juice of bitter gourd mixed with a teaspoonful of lime juice should be taken, sip by sip, on empty stomach daily for four to six months in these conditions. Its regular use in endemic regions of leprosy acts as a preventive medicine.

Respiratory Disorders

Bitter gourd plant roots are used in folk medicine for respiratory disorders from ancient times. A teaspoonful of the root paste mixed with equal amount of honey or Tulsi leaf juice, given once every night for a month acts as an excellent medicine for asthma, bronchitis, Pharyngitis, colds and Rhinitis.

Alcoholism

Leaf juice is beneficial in the treatment of alcoholism. It is an antidote for alcohol intoxication. It is also useful in liver dam age due to alcoholism. It is now karela, jamun & brinjal: Biopiracy of rich Indian herbal wealth. Earlier R Basmati rice had been patented by Rice Tec Inc of USA. Now karela (bitter gourd), jamun (blackberry), gurmar, and brinjal have been patented by a MNC in the USA. The Supreme Court has issued notice to the Union Agriculture Ministry as the petitioner, Research Foundation for Science, Technology and Ecology (RFSTE), charged the Centre with failure to protect the country's biodiversity despite giving an assurance.

RFSTE counsel Sanjay Parikh said Attorney-General Soli Sorabjee had informed in 1998 about action taken by the government in protecting the biodiversity by bringing biodiversity legislation. As far as basmati rice was concerned, Sorabjee had said the government had already taken steps to challenge the grant of patent, but there has been no follow-up after that. Biopiracy is an epidemic. Earlier, neem, haldi, pepper, harad, bahera, amla, mustard, Basmati, ginger, castor, jaramla, amaltas, isabgol, and now karela and jamun have been patented under the USIPR system.

A patent number US6,900.240 was granted recently to Cromak Research Inc based in New Jersey, on edible herbal compositions for anti-diabetic properties. It comprised mixtures of at least two Indian herbs selected from a group consisting of 'syzygium jambolanum cumini', popularly known as jamun, 'momordica charantia' (bitter gourd or karela); 'solanum melongena' (brinjal or egg plant' and 'gymaema sylvestre' (gurmar) as anti-diabetes agents for their proposed use in reducing sugar.

Patents had been granted on May 4 last in the USA on edible herbal composition comprising mixtures of herbs selected from the group consisting of jamun, gurmar, karela and brinjal useful as two hypolycemic agents. The investors include two non-resident Indians Onkar S. Tomar and Kirpanath Borah along with their American colleague Peter Glomski. The patenting of these anti-diabetic plants has again highlighted the problem of biopiracy of rich Indian herbal wealth.

A patent issued in the USA does not affect us dramatically, says the Director-General of the Council of Scientific and Industrial Research (CSIR), Dr R.A. Mashelkar. "It does not mean that one cannot use similar mixture in India for anti-diabetic treatment. Yes, it may affect the possibility of our exporting such a mixture to the USA", says Mashelkar. India should be more active in filing the patents. "Since 1994 when the TRIPS came into force, the US in the last four years has granted upto 1,890 patents. Most of these have been from China. India's contribution has been meagre," says Mashelkar.

Jamun belongs to the guava family. It originated in India and is now naturalised throughout the Far East countries. It is a fruit tree of considerable economic value, says K.V. Peter, Director, Indian Institute of Spice Research, Kozhikode. Extracts of stems, leaves, buds and flowers possess moderate antibiotic activity against micrococcus pyogenous, aureus, according to "Wealth of India", a CSIR publication. Experiments conducted at the Central Drug Research Institute, Lucknow, show that oral administration of dried alcoholic extracts of seeds to diabetic patients reduces the blood sugar level.

The government is yet to take a decision on contesting patents obtained by the US firm for kerala, jamun, brinjal and gurmur, because it is still examining and analysing the whole issue. While the fruit, leaves and roots of bitter gourd have long been used in India as a folk remedy for diabetes mellitus, the leaves of gurmar are useful in the management of maturity onset diabetes. It is an important ingredient in Ayurvedic formulation for diabetes. Their use in the treatment of diabetes is documented in the authoritative treatises such as "Wealth of India", "Compendium of Indian Medicinal Plants" and the "Treatise on Indian Medicinal Plants". The patent document has not mentioned the above findings under the 'prior art' states the Intellectual Property Rights (IPR) bulletin published by Technology Information Forecasting and Assessment Council (TIFAC).

With 70 per cent of the country's population depending on non-allopathic medicines, the potential of traditional medication is large. This increases the importance of ensuring that the traditional knowledge base of the country is protected and that the multinational companies are not allowed to patent traditional medicines says the former Controller-General, Patent, Designs and Trademarks, Mr K.V. Swaminathan.

The problem of biopiracy is a result of Western style of IPR systems, and not the absence of such IPR systems in India. Therefore the implementation of Trade Related Intellectual Property Rights (TRIPS) Agreement which is based on the US style patent regime must immediately be stopped. The promotion of piracy is not an aberration in the US patent law. It is intrinsic to it. Article 102 of the US Patent Law, which defines 'prior art', does not recognise technologies and methods in use in other countries as 'prior art'. If knowledge is new for the US, it is novel, even if it is part of an ancient tradition of other cultures of countries.

1. US Patent on Neem
2. Development-India: Southern States Takes on biopiracy
3. Tribe Accuses Biologists
4. Hoodia cactus: Western drug industry exploits developing countries
5. Pelargonium reniforme Effective in Treating Bronchitis, a South African plant: The Rape of the Pelargoniums

Patents Bill: Protecting Indigenous Knowledge (IK) but not enough to provide for the protection of IK

The Patents Amendment Bill is the third in the series of amendments that were undertaken to make the Patent Act 1970 conform to India's full obligations under TRIPS by 2005. The patent bill makes a number of changes, such as removing exclusions of food and medicines from being patentable, introduction of representation as a mechanism for opposition and compulsory licensing.

The import of the changes that have been sought to be made to the Patent Act through the bill should be seen in the context of the overt recognition given by the state to the importance of protecting indigenous knowledge (IK). The state has finally woken up to the realisation that only through extending legal recognition to IK (Substantially under the Biological Diversity Act 2002) it can be protected from usurpation and unfair exploitation. Legal recognition of IK (at least in principle) entails its identification as a separate knowledge system, the recognition that it is intimately weaved into the livelihood and culture of indigenous communities, and understanding its dialectical relationship with the surrounding biodiversity.

Though the Patent Bill has seemingly provided for the protection of IK (by putting into place mechanisms for the tracing of IK inputs in an 'invention'), it leaves open several loopholes through which biopiracy and usurpation of IK could be easily practised. Hence it is just not enough to provide for the protection of IK by introducing a single clause prohibiting patents derived from IK. It is imperative to realise that only by way of a holistic integration of the two objectives of protection of IK and granting patent rights, could there be a realisation of the primary objective of IK protection in real terms. The Patents Amendment Bill does not go far enough to develop a holistic framework for the protection of indigenous knowledge.

It thus leaves several loopholes through which biopiracy and usurpation of indigenous knowledge can easily take place. A more synergistic relationship between the patent authority and the National Biodiversity Authority, and a more transparent and inclusive decision-making process, will help remove some of the shortcomings in the bill.

The mere inclusion of such a general clause is not sufficient to protect larger social objectives and goals, especially that of protection of indigenous knowledge. This could be done in two steps.

Firstly, it should be expressly stated that the violation of any listed principles should be a ground for revocation or compulsory licensing. This would make the provision enforceable and thus inherently enabling.

Secondly, the phrase 'public health' should be extended to include the right to access biological resources and health remedies sourced from therein. And therefore if the grant of a patent right over certain resources (micro-organisms being patentable) circumscribes this right of local or indigenous communities, it should be interpreted to mean violation of public health needs (N. Choudhury, EPW, November 20, 2004).

Plant Genetic Resources for Food and Agriculture

After prolonged negotiations the Treaty on Plant Genetic Resources for Food and Agriculture (IPGR) came into being in November, 2001.

For the purposes of the treaty, the term 'plant genetic resources' (PGR) shall mean any genetic material of plant origin of actual or potential value for food and agriculture. 'Genetic material' under the Treaty means any material of plant origin, including reproductive and vegetative propagating material, containing functional units of heredity. The IPGR was approved during the FAO Conference (31st Session, resolution 3/2001) in November 2001, with 116 votes and 2 abstentions (USA and Japan).

The Treaty has two main objectives (Article 1 of the Treaty):

- conservation and sustainable use of plant genetic resources for food and agriculture
- fair and equitable sharing of the benefits arising out of their use for sustainable agriculture and food security.

After prolonged negotiations the Treaty on Plant Genetic Resources for Food and Agriculture (IPGR) came into being in November, 2001. The present version of the Treaty has deviated from the International Undertaking on Plant Genetic Resources for Food and Agriculture agreed in 1981 that relied on the principle of genetic resources being common heritage of humanity. Instead, relying on the Convention on Biological Diversity (CBD) the IPGR has brought genetic resources under the jurisdiction of national governments. According to many, the Treaty has left some of the central issues unresolved while some of its provisions are open to interpretation. The points raised by them include:

- The list of food crops, forage and their relatives included in the treaty is not exhaustive (soya, sugar cane, oil palm and groundnut etc.)
- The extent intellectual property rights will be allowed within the treaty rules is unclear
- The extent to which farmers and communities will be allowed to freely use, exchange and breed the seeds is unclear
- The mechanism for sharing benefits over the commercial use of genetic material in terms of amount, form and conditions remain unclear
- Enforcement procedure to be used by national governments for ensuring compliance is not detailed out.

Moringa Oleifera

There are fourteen known species of trees belonging to the genus Moringaceae. Moringa stenopetala is native to Ethiopia and northern Kenya. M. peregrina is found in the Sudan, Egypt, the Arabian peninsula and as far north as the Dead Sea. M. ovalifolia grows in Angola and Namibia. However, the best known member of the genus is Moringa oleifera, a fast-growing, drought-resistant tree native to sub-Himalayan tracts of northern India but now distributed world-wide in the tropics and sub-tropics. The pleasant-tasting edible oil can be extracted from the seeds for use in making perfume, protecting skin, and for use as a lubricant for fine machinery. Powder from seed kernels works as a natural coagulant which can clarify even very turbid water, removing up to 99% of the bacteria in the process. Moringa trees are also well-suited for use in alley cropping systems, and various parts and products can be used as animal forage, a domestic cleaning agent, dye, fertilizer, honey and sugar cane juice clarifier, honey producer, live fencing, traditional medicine, ornamental, plant disease preventative, paper pulp, rope fiber, and hide tanning agent.

Moringa Oleifera Lam.

This species is one of the world's most useful plants. Though apparently native only to restricted areas in the southern foothills of the Himalayas, M. oleifera is cultivated in all the countries of the tropics. M. oleifera is cultivated for its leaves, fruits, and roots for a variety of food and medicinal purposes. The young fruits (sometimes called "drumsticks") can be cooked in a number of different ways. An excellent oil is derived from the seeds, which is used for cooking and lubrication of delicate mechanisms. The leaves are extensively used as a vegetable in many parts of the world, and the root can be made into a condiment similar to horseradish (true horseradish, Armoracia rusticana, is a member of the Mustard Family, Brassicaceae). M. oleifera is also of interest because of its production of compounds with antibiotic activity such as the glucosinolate 4 alpha-L-rhamnosyloxy benzyl isothiocyanate. Other research has focused on the use of M. oleifera seeds and fruits in water purification.

MORINGA: A MEDICAL PHARMACOPOEIA

Moringa oleifera, commonly referred to simply as Moringa, is the most widely cultivated variety of the genus Moringa. It is of the family Moringaceae. It is an exceptionally nutritious vegetable tree with a variety of potential uses. The tree itself is rather slender with drooping branches that grows to approximately 10 m in height; however, it normally is cut back annually to one meter or less, and allowed to regrow, so that pods and leaves remain within arms reach.

The Moringa tree grows mainly in semi-arid tropical and subtropical areas. While it grows best in dry sandy soil, it tolerates poor soil, including coastal areas. It is a fast-growing, drought-resistant tree that apparently is native only to the southern foothills of the Himalayas. Today it is widely cultivated in Africa, Central and South America, Sri Lanka, India, Mexico, Malaysia and the Philippines. Considered one of the world's most useful trees, as almost every part of the Moringa tree can be used for food, or has some other beneficial property. In the tropics it is used as foliage for livestock.

The leaves are highly nutritious, being a significant source of beta-carotene, Vitamin C, protein, iron and potassium. The leaves are cooked and used as spinach. In addition to being used fresh as a substitute for spinach, its leaves are commonly dried and crushed into a powder, and used in soups and sauces. The seeds may be crushed and used as a flocculant to purify water. The Moringa seeds yield 38-40% edible oil (called Ben oil, from the high concentration of behenic acid contained in the oil) that can be used in cooking, cosmetics, and lubrication. The refined oil is clear, odorless, and resists rancidity at least as well as any other botanical oil. The seed cake remaining after oil extraction may be used as a fertilizer.

Delicacy in West Bengal and Bangladesh

The flowers are also cooked and relished as a delicacy in West Bengal and Bangladesh, especially during early spring. There it is called Sojne ful and is usually cooked with green peas and potato.

In South India, it is used to prepare a variety of sambar and is also fried. In Tamil it is called Murungakai. In Gujarati is it called Saragvo. It is also preserved by canning and exported worldwide. In other parts of India, especially West Bengal and also in a neighboring country like Bangladesh it is enjoyed very much. It can be made into varieties of curry by mixing with coconut, poppy seeds and mustard. It can just be boiled, until the drumsticks are semi-soft and consumed directly without any extra processing or cooking. It is used in curries, sambars, kormas, and dals, although it is also used to add flavor to cutlets, etc. Tender drumstick leaves, finely chopped, make an excellent garnish for any vegetable dishes, dals, sambars, salads, etc. One can use the same in place of or with coriander, as these leaves have high medicinal value. If the pulp has to be scraped out after cooking the sticks, then keep the pieces as long as 4-5 inches long. Also do not scrape the skin before boiling. This will help to hold and scrape them more easily and with less mess. For drumstick sambar follow recipe for traditional sambar, adding boiled drumstick fingers, along with onions in the oil, while stir frying.

Drumstick dal, is also a very tasty version of the traditional 'toor dal'. Add some of the pulp to the boiled dal, and hand beat it along with the dal before seasoning. This will give an unusual, novel flavor to this dal. In another variation you may add pieces of boiled drumstick including the water in which it was boiled, to the traditional toor dal while it is simmering. The pieces are delightful to chew on with the dal and rice.

M.oleifera in Water Treatment

Sanskrit writings in India dating from several centuries BC make reference to seeds of the tree Strychnos potatorum as a clarifier, Peruvian texts from the 16th and 17th centuries detail the use by sailors of powdered, roasted grains of Zea mays as a means of settling impurities. More recently, Chilean folklore texts from the 19th century refer to water clarification using the sap from the 'tuna' cactus (Opuntia fiscus indica). However, of all the plant materials that have been investigated over the years, the seeds from M.oleifera have been shown to be one of the most effective as a primary coagulant for water treatment.

The traditional use of the M.oleifera seeds for domestic household water treatment has been limited to certain rural areas in the Sudan. Village women collecting their water from the River Nile would place powdered seeds in a small cloth bag to which a thread is attached. This would then be swirled around in the turbid water. Water soluble proteins released from the powdered seeds, attach themselves to, and bind between, the suspended particles forming larger, agglomerated solids. These flocculated solids would then be allowed to settle prior to boiling and subsequent consumption of the water. Since the early 1970's a number of studies have been carried out to determine the effectiveness of the seeds for the treatment of surface water at individual household level. Utilising artificially prepared turbid water and naturally turbid raw waters, laboratory investigations have confirmed the seeds to highly effective in the removal of suspended solids from waters containing medium to high initial turbidities.

At low turbidities, as may be experienced during the dry season, the seeds are less effective although their performance is very much dependant on the raw water to be treated. Work is currently underway at the University of Leicester examining the potential of utilising the seeds within a contact flocculation filtration process for the treatment of low turbidity water. Preliminary results have demonstrated some considerable success.

All Moringa food products have a very high nutritional value. You can eat the leaves, especially young shoots, young pods, flowers, roots, and in some species even the bark. Leaves are low in fats and carbohydrates and rich in minerals, iron and vitamin B.

According to Verma et al. (1976), "saijan" is a fast growing tree being planted in India on a large scale as a potential source of wood for the paper industry. It seems doubtful that the wood and seed oil could both be viewed as fountains of energy. According to Burkill (1966), "The seeds yield a clear inodorous oil to the extent of 22 to 38.5 percent. It burns with a clear light and without smoke. It is an excellent salad oil, and gives a good soap... It can be used for oiling machinery, and indeed has a reputation for this purpose as watch oil, but is now superseded by sperm oil." Sharing rather similar habitat requirements with the jojoba under certain circumstances, it might be investigated as a substitute for sperm whale oil like jojoba. Growing readily from cuttings, the ben oil could be readily produced where jojoba grows.

Root-bark yields two alkaloids: moringine and moringinine. Moringinine acts as cardiac stimulant, produces rise of blood-pressure, acts on sympathetic nerve-endings as well as smooth muscles all over the body, and depresses the sympathetic motor fibers of vessels in large doses only.

Uses: Almost every part of plant is of value for food. Seed is said to be eaten like a peanut in Malaya. Thickened root used as substitute for horseradish.

Foliage eaten as greens, in salads, in vegetable curries, as pickles and for seasoning. Leaves pounded up and used for scrubbing utensils and for cleaning walls. Seeds yield 38-40% of a non-drying oil, known as Ben Oil, used in arts and for lubricating watches and other delicate machinery. Oil is clear, sweet and odorless, never becoming rancid; consequently it is edible and useful in the manufacture of perfumes and hairdressings. Wood yields blue dye. Leaves and young branches are relished by livestock. Commonly planted in Africa as a living fence (Hausa) tree. Trees planted on graves are believed to keep away hyenas and its branches are used as charms against witchcraft. Bark can serve for tanning; it also yields a coarse fiber.

Folk Medicine: According to Hartwell (1967-1971, Hartwell, J.L. 1967-1971. Plants used against cancer. A survey. Lloydia 30-34.), the flowers, leaves, and roots are used in folk remedies for tumors, the seed for abdominal tumors. The root decoction is used in Nicaragua for dropsy. Root juice is applied externally as rubefacient or counter-irritant. Leaves applied as poultice to sores, rubbed on the temples for headaches, and said to have purgative properties. Bark, leaves and roots are acrid and pungent, and are taken to promote digestion. Oil is somewhat dangerous if taken internally, but is applied externally for skin diseases. Bark regarded as antiscorbic, and exudes a reddish gum with properties of tragacanth; sometimes used for diarrhea. Roots are bitter, act as a tonic to the body and lungs, and are emmenagogue, expectorant, mild diuretic and stimulant in paralytic afflictions, epilepsy and hysteria.

Content	Mohringa	other food
Vitamin A	6,780 mg	caroot: 1,890 mg
Vitamin C	220 mg	Orange: 30 mg
calciumm	440 mg	cow milk: 120 mg
potassium	259 mg	Banana: 88 mg
Protein	6. 6 g	cow milk: 3,2 g

Many of the above vitamins, minerals and amino acids are very important for a healthy diet. An individual needs sufficient levels of certain vitamins, minerals, proteins and other nutrients for his physical development and well-being. A deficiency of any one of these nutrients can lead to health problems. Some of the problems caused by deficient diets are well known: scurvy, caused by lack of vitamin C; night blindness, caused by lack of vitamin A; kwashiorkor, caused by lack of protein; anemia, caused by lack of iron. Many other health problems are caused by lack of vitamins or minerals which are less known, but still essential to a person's bodily functions.

Besides it contains in pods and leaves:

	Pods	leaves
Vitamin A-B carotene (mg) 0.1 6.8	0.1	6.8
Vitamin B-choline (mg)	423	423
Vitamin B1-thiamin (mg)	0.05	0.21
Vitamin B2-riboflavin (mg)	0.07	0.05
Vitamin B3-nicotinic acid (mg)	0.2	0.8
Arginine (g/16g N)	3.6	6.0
Histidine (g/16g N)	1.1	2.1
Phenylanaline (g/16g N)	4.3	6.4
Leucine (g/16g N)	6.5	9.3
Isoleucine (g/16g N)	4.4	6.3
Valine (g/16g N)	5.4	7.1

Source: Church World Service,Dakar, Senegal, 1999.

Actual need for different vitamins, etc., will vary depending on an individual's metabolism, age, sex, occupation and where he/she is living. Recommendations for daily allowances (RDA) also vary according to whom is doing the study. WHO/FAO recommend the following daily allowances for a child aged 1-3 and a woman during lactation

Leaves and pods of Moringa oleifera can be an extremely valuable source of nutrition for people of all ages. For a child aged 1-3, a 100 gram serving of fresh leaves would provide all his daily requirements of calcium, about 75% of his iron and half his protein needs, as well as important supplies of potassium, B complex vitamins, copper and all the essential amino acids. As little as 20 grams of fresh leaves would provide a child with all the vitamins A and C he needs. Iron is a vital component of red blood cells which carry oxygen. Iron assists the muscles to keep reservoirs of oxygen and makes the body more resistant to infections. Iron deficiency can cause anemia, tiredness, headaches, insomnia and palpitations. In children, deficiency can cause slow growth and impaired mental performance. Fresh Moringa leaves contain over three times the amount of iron found in spinach (2.1mg/100g).

Sulfur is a constituent of all proteins and an essential element for all life. In the body, the sulfur content is mostly found in the skin, joints, nails and hair. The more sulfur content in the hair, the curlier it will be (sheep hair is about 5% sulfur). Although involved in many metabolic processes, there is generally not a recommended dietary requirement for sulfur because the body can extract it from the amino acids cysteine and methionine. An acid also found in strawberries, rhubarb and spinach, oxalic acid can combine with calcium and iron in the body to form insoluble compounds which the body cannot absorb. However, only large amounts of oxalic acid consumption are liable to cause calcium and iron deficiencies.

For pregnant and breast-feeding women, Moringa leaves and pods can do much to preserve the mother's health and pass on strength to the fetus or nursing child. One portion of leaves could provide a woman with over a third of her daily need of calcium and give her important quantities of iron, protein, copper, sulfur and B vitamins. Just 20 grams of fresh leaves will satisfy all her daily requirement of vitamin C. For both infants and mothers, pods can be an important source of fiber, potassium, copper, iron, choline, vitamin C and all the essential amino acids.

Malnourished children can benefit from addition of Moringa leaves to their diet. The high concentrations of iron, protein, copper, various vitamins and essential amino acids present in Moringa leaves make them a virtually ideal nutritional supplement.

Moringa leaves can be dried and made into a powder by rubbing them over a sieve. Drying should be done indoors and the leaf powder stored in an opaque, well-sealed plastic container since sunlight will destroy vitamin A. It is estimated that only 20-40% of vitamin A content will be retained if leaves are dried under direct sunlight, but that 50-70% will be retained if leaves are dried in the shade.

This powder can be used in place of fresh leaves to make leaf sauces, or a few spoonfuls of the powder can be added to other sauces just before serving. Addition of small amounts of leaf powder will have no discernible effect on the taste of a sauce. In this way, Moringa leaves will be readily available to improve nutritional intake on a daily basis. One rounded soup spoon of leaf powder will satisfy about 14% of the protein, 40% of the calcium, 23% of the iron and nearly all the vitamin A needs for a child aged one to three. Six rounded spoonfuls of leaf powder will satisfy nearly all of a woman's daily iron and calcium needs during times of pregnancy and breast-feeding (Church World Service, 2000).

Haripada's Haridhan (Haririce): A Farmer's Pride

During the current season, all the lands of Jhenidah district have been filled with Haridhan. The Agricultural department of the Government after examining the paddy has declared Haridhan a profitable cultivation. It costs little to cultivate it. Moreover, compared to the profuse growth of it, the expenditure for its cultivation seems to be very little. The farmers are getting bumper crop from cultivating Haridhan. They get 18 to 20 maunds paddy per bigha. This year, too, they expect to have bumper production.

Haripada Kapali, the only son of a poor farmer, lost his father at the age of 8. Instead of going to school, he was burdened with the responsibility of running a family. Since then he has been working as a farmer. He is now 75 years old. Being almost illiterate, he produces good crops using his intelligence only. He has a good reputation in the locality for his skill in growing all kinds of crops. Even at the fag end of his life he has invented a super quality paddy with the help of his instinctive knack for farming. This special kind of paddy is known as Haridhan in Jhenidah.

Haridhan has been grown sufficiently in the last 7 consecutive seasons. 4 thousand hectors land from only 4 Satak have been brought under Haridhan cultivation. During the current season, all the lands of Jhenidah district have been filled with Haridhan. The Agricultural department of the Government after examining the paddy has declared Haridhan a profitable cultivation. It costs little to cultivate it. Moreover, compared to the profuse growth of it, the expenditure for its cultivation seems to be very little. The farmers are getting bumper crop from cultivating Haridhan. They get 18 to 20 maunds paddy per bigha. This year, too, they expect to have bumper production. On 20 October, 2006, I went to pay a short visit to Haripada Kapali at his home in Ashan Nagar to find out about him and the paddy he invented, how he was doing, how much the cultivation of Haridhan was spreading, how much it would grow this year etc.

Taking a long time Haripada narrated the past and present experiences of his farmer's life. At one point he got a little emotional and said, "If the Government recognises my contribution as you and the press have done with regard to my invention, one day this Haridhan will spread all over the country and I will live in the midst of Haridhan."

Haripada Kapali comes of a poor family of Enayetpur village in Jhenidah Sadar. He is the only son of Ishwar Kundu. His father died when Hari was eight years old. His father left behind 5 bighas of land but his uncles misappropriated 3 bighas of land playing tricks on him who was just a child at that time. He started cultivation on the remaining two bighas of land immediately in order to maintain his family. Meanwhile, he finished his elementary education. Haripada recalled that crops did not grow well in those days. They had to pass their days through great hardship. He kept himself busy in agricultural work round the clock. Haripada has no offspring.

How Haridhan was Invented

Haripada Kapali told us the story of how he discovered Haridhan and started producing it. About seven or eight years ago, while weeding out a plot of B.R-11 paddy one day, he came across a bunch of paddy which was different from the other paddy. This bunch of paddy was more beautiful to look at and taller, too. He took special care of it and facilitated its growth in a natural way. At first he thought this would not yield good crops. But finally, this separate paddy gave better production than the already planted ones. This new kind of paddy plants was stout and taller. Haripada collected this paddy separately and stored it in a safe place. Next year he made separate seeds, beds and sowed this 2/3gram seeds in another plot. He took care of both the paddy equally. At last, after harvesting he estimated that Haridhan grew more abundantly in comparison with the planted ones. He preserved all the seeds and planted 16 satak land with these seeds the next year.

Haripada informed us that when the paddy grew on the sixteen Satak land, the nearby villagers rushed to see the profuse yield of crops with their own eyes. People became highly pleased to see the bumper production of Haridhan, the new paddy they named after Haripada. Later, villagers from all the neighbouring places assembled in Haripada's house to collect the seeds. Haripada said that many of them were dissatisfied, as he couldn't provide them with the seeds. Thus haridhan is being cultivated in this locality for the last seven or eight years.

Farmer Sukur Ali of Ashan Nagar village says that out of 10 bighas of paddy that he has planted this year, 5 bighas are Haridhan. It costs less to grow than what it does for the other paddy, but yields more crops. He has harvested 12/13 maunds Swarna paddy per bigha, whereas, he has collected 18/20 maund Haridhan per bigha. Adhir Kumar of the same village informs that he has two bighas of land that have been brought under Haridhan cultivation. He adds by saying that those who cultivated other paddy have sustained a loss.

Local Public Representative's Demand

Mizanur Rahman, U.P Chairman of Madhuhati Union where Haridhan was first cultivated says that it needs no telling that Haridhan is a paddy of super quality. As it gives a bumper harvst, its cultivation is increasing every year. The Chairman also informs that rice researchers have already collected this Haridhan from Haripada. He demands that this paddy, known as "Haridhan" in his locality, should be recognized officially (Azibor Rahman, Novenber, 2006).

SPICES AND HERBS

In Indian subcontinnent, the reference to the curative properties of some herbs in the Rig Veda seems to be the earliest records of use of plant in medicine. The period of Rig Veda is estimated to be between 3500 and 1800 B. C.

Empires have been settled and unsettled by the spice and herb trade. Lowenfed and Back in their book "Herbs and Spices" describe: "How did the Phoenicians suddenly become so powerful, and how could even a small city like Venice employ an army with such a strategist as Othelo? What made it possible for the Dutch people, being a small nation, to develop at one time a great empire? Why were Vasco da Gama and Christopher Columbus so interested to find a new sea route to India? How could it happen that the British East India Company became a political power? The answer is spices, spices and again spices."

The London Times comments (The Thing from outer Spice): "What first led Europeans to spread all over the globe? Was it religion and the rise of capitalism? Or had it more to do with pepper, which was essential to mask the flavour of salted meat, stinking fish and boring vegetable."

Richard Evans Schultes wrote in early 60s: Civilisation is on the march in many, if not most, primitive regions. It has long been on the advance, but its space is now accelerated as the result of world wars, extended commercial interests, increased missionary activity, widened tourism. The rapid divorcement of primitive peoples from dependence upon their immediate environment for the necessities and amenities of life has been set in motion, and nothing will check it now. One of the first aspects of primitive culture to fall before the onslaught of civilisation is knowledge and use of plants for medicines. The rapidity of this disintegration is frightening Our challenge is to salvage some of the native medico-botanical lore before it becomes forever entombed with the cultures that gave it birth.

In recent decades science has 'rediscovered' what 'primitive peoples' intuitively understood: namely, that all living organisms profoundly interact both with one another and with their non-living surroundings. The modern study of this system of myriad interactions is called ecology. Recent research carried out in China indicates that some plants are ableto suppress the AIDS virus. For example bitter gourd. This plant belongs to the Lagenariaceae family. Its root, leaf, fruit and seed have been found to exhibit medicinal values.The plant is widely found in southern China. The Chinese believe that the root, which is bitter in taste, is cold in natureand can cure fever and remove toxinsfrom the body.

In ancient times people did not differentiate between food and medicine. They believed that food was both diet and medicine. Because of their therapeutic value, the use of major spices, curry and herbs in itself is a protection against spoilage and contamination while aiding digestion. Ginger prevents dyspepsia, garlic controls cholesterol and hypertension, onions, fenugreek, mint and pepper are germifuges and often act as anti-histamines. For thousands of years turmeric has beens used as a stimulant in native medicine-often administered in disorders of blood. It is widely used as an external applicant in bruises, leach bites etc. The Indigenous system of medicine has given an extra special place to spices because of their unique medicinal properties. Clove oil is applied to relieve toothaches. Pepper added to hot tea is a patent "grandma's treatment" for common cold. Turmeric has anti-bacterial properties and its solution can be used as an antiseptic for cleaning wounds.There are literally thousands of medicinal uses for such spices. Even today in much of rural India, the wisest doctors are often the mothers and grandmothers who know the uses of their "kitchen pharmacies." Ajwain (Carum copticum) is used as a medicine in stomach ailments. It has stimulant, tonic and carminative properties and the anti-spasmodic virtues of asafoetida.

Medicinal Properties:

- Loss of Appetite: Mix and powder equal quantities of ajwain,saunf,ginger and salt.Mix a teaspoon of this mixture in boiled tice along with ghee and eat thrice a day.
- Colic Pains,Indigestion,Gas:Grind 2 tsp each ajwain and dried ginger into a fine powder.Add a little black salt.Take 1 tsp of this mixture with 1 teacup warm water frequently.

- Kidney-pain,renal colic: Mix and grind 1 tbsp black cumin 2 tsp ajwain and 1 tsp black salt into a fine powder.Add 1 tsp brown vinegar.Take 1 tsp of this mixture every hour till symptoms subside.
- Nasal congestion in children:Crush a fistful of ajwain and tie up in a cotton napkin and place it near the pillow.
- Common Cold,Congestion in the chest: Boil 1/2 tsp ajwain along with 1 pinch of turmeric powder, in half a cup of water.Cool.Add 1 tsp honey and drink.Inhale vapours of ajwain boiling in a pan of water.
- Cough: Mix 1/2 tsp ajwain seeds,2 cloves and a pinch of salt.Powder and sip with a little warm water frequently.
- Respiratory problem due to blockage of dried phlegm: Crush 2 tsp ajwain seeds.Mix in a glass of buttermilk and drink.

Arjun Terminalia arjuna its bark is astrinent and is used in fevers and in fractures and contusions; it is also taken as a cardiac tonic.

Clinical evaluation of this botanical medicine indicates it can be of benefit in the treatment of coronary artery disease, heart failure, and possibly hypercholesterolemia. It has also been found to be antibacterial and antimutagenic. Terminalia's active constituents include tannins, triterpenoid saponins (arjunic acid, arjunolic acid, arjungenin, arjunglycosides), flavonoids (arjunone, arjunolone, luteolin), gallic acid, ellagic acid, oligomeric proanthocyanidins (OPCs), phytosterols, calcium, magnesium, zinc, and copper.1,2.

A clinical trial using 500 mg of an extract tid for DCM (Dilated Congestive Cardiomyopathy, the most common type of cardiomyopathy). Patients with severe heart failure showed improvement in heart function within 2 weeks and improvement which continued for the following 2 years. The arjun in this trial was concentrated, but not standardized, as are some commercial preparations (1% arjunolic acid). (Int J Cardiol 1995;49: pp.191-9).

Traditional Ayurvedic Uses:

- Arjuna is a very large tree. The bark is used in certain herbal combinations as a powerful, soothing tonic for the heart. It is good for both the physical heart as a muscle, as well as for the emotions associated with the heart.
- Arjuna is used for loneliness, sadness and frustration. It strengthens the emotions to decrease excessive response to stress and trauma.
- It helps strengthen the body's natural rejuvenative processes, hastening the replacement of dead or weak cells with fresh, vital ones. In proper combinations, Arjuna helps stabilize an erratic heart beat.
- Arjuna helps balance all three doshas at once (Vata, Pitta, Kapha), a rare and very valuable property.
- This tree herb bears the same name as Arjuna, son of Pandu--a great hero of the Bhagavad-Gita. The Gita is a treasured poem from the Vedic epic called the Mahabharata.

Asafoetida Ferula asafetida Linn: Asafoetida, the gum resin prized as a condiment in India and Iran, is obtained chiefly from plant Ferula asafetida. The Latin name ferula means "carrier" or "vehicle". Asa is a latinized form of Farsi asa "resin ", and Latin foetidus means "smelling, fetid". In ancient Rome, asafoetida was stored in jars together with pine nuts, which were alone used to flavour delicate dishes. Another method is dissolving asafoetida in hot oil and adding the oil drop by drop to the food. If used with sufficient moderation, asafoetida enhances mushroom and vegetable dishes, but can also be used to give fried or barbecued meat a unique flavour. Asafoetida is a useful antidote for flatulence. There are claims for it being used to cure bronchitis and even hysteria.

Basil (tulsi) Ocimum sanctum Bengali Tulsi, Hindu poets say that it protects from misfortune and sacrifices and guides to heaven all who cultivate it. The leaves have expectorant properties, and their juice is used by native physicians for catarrh and bronchitis. This preparation is also applied to the skin in ring-worm and other cutaneous diseases. The infusion of the leaves is used for gastric disorders and hepatic affilation. Leaves are said to have diaphoretic properties. The oil obtained from leaves has the property of destroying bacteria and insects. The juice or infusion of the leaves is useful in bronchitis, catarrh, digestive complaints; it is applied locally on ringworm and other skin diseases.

Bay leaf Laurus nobilis leaf of the sweet bay tree. Is an evergreen plant, indigenous to Asia Minor bordering the Mediterranean. Bay is a tree of the sun under the celestial sign of Leo and has been cultivated from ancient times; its leaves constituted the wreaths of laurel that crowned emperors, heroes and victorious athletes in ancient Greece and Rome.Bay leaves contain approximately 1.5-2.5 % essential oil, the principal component of which is cineole. Bay oleoresin contains about 4-8 % volatile oil. Bay leaves are a popular culinary flavouring in classic and contemporary cuisines which stimulates the appetite. Bay leaf has legendary medicinal properties. It has astringent, diuretic and digestive qualities and is a good appetite stimulant.

Bel Aegle Marmelos (L), family: Rutaceae Bengali Bel fruit is valuable chiefly for its mucilage and pectin; it is very useful in chronic diarrhoea and dysentry. The antibiotic activity of the leaf, fruit and root of this plant has been confirmed.

Cardamom Zingiberaceae Bengali Elachi: is used chiefly for relieving flatulence or feeling of overfullness of stomach, i. e. to promote digestion. It is administered with purgatives, and as a flavour agent.

Digitalis Digitalis Purpurea L. Hindi Tilpusphi the dried leaves of the plant constitute the drug. The main use of this drug is in heart diseases. The drug promotes and stimulates the activity of all muscle tissues. The drug promotes the activity of all muscle tissues. It is used in cases of congested heart failure. Digitalis forces more blood into the coronaries and improves the nutrition. It improves the blood supply to the kidney and this promotes urination, and removes obstruction in kidneys. It is used in some ointment for local application on wounds and burns. All chile peppers come from the Americas. Columbus "discovered" them in the West Indies and brought them back to Europe where they eventually spread to the rest of the world. The burning sensation from eating chile peppers is caused by a group of compounds called capsaicinoids. It is often simply called capsaicin, though this is simply one of the capsaicinoids. These compounds are concentrated in the white veins in the pepper that hold the seeds.

The reason is that during the eating of chillies, a chemical in the chillie pepper called Capsaicin, irritates the trigeminal cells. These are pain receptor cells located throughout the mouth, the nose and the throat. When your body's nerves feel the pain induced by the chemical on these cells, they immediately start to transmit pain messages to your brain. Your brain receives these signals and responds by automatically releasing endorphins (the body's natural painkiller). These endorphins kick in and act as a painkiller and at the same time, create a temporary feeling of euphoria, giving the chillie pepper eater, a natural high.

Medicinal, Pharmacological and Biochemical Properties

Pain relief, especially for arthritis and joint pain, is the most common usage right now. Many creams for pain relief now contain capsaicin. The depletion of substance P in the nerves help to reduce nagging pain. Another medical use, amazingly enough, is in the treatment of ulcers. Since the discovery of the bacteria, Helicobacter pylori, as the primary cause of ulcers, antibiotics have been the common treatment. Chile peppers have natural antibiotic properties. As well, they stimulate the mucosa of the stomach.. Low in calories, peppers contain twice as much vitamin C, per weight, as citrus fruits and more vitamin A than carrots (especially red chiles). As well, peppers aid in digestion and speed up metabolism. The capsaicin of chillies does have a medical application, in the truest sense of the word. It is used in plasters to be applied externally in cases of severe muscle pain, acting in much the same way as the pleasantly 'hot' menthol creams. Internally, chillies and all members of the capsicum family are rich in vitamin C. They are reputed to help keep capillaries from hardening, thus lessening the risk of cardiovascular disease. The irritant effect of chilli peppers is used as a mask for pain from conditions such as rheumatism and nerve pain. Medicines such as these are known as counter-irritants and there are several containing extracts of chilli.

A study published in the British Journal of Nutrition has discovered that capsaicin, when added to breakfast foods or appetizers at lunch, causes people to eat less during meals and for hours afterwards. Thirteen women, who ate breakfast foods spiced with red pepper, ate less than normal at breakfast and during the day, while ten men, who ate red pepper laced appetizers, consumed fewer calories at lunch and during a mid-day snack hours later. Aside from acting as an appetite suppressant, red pepper also seems to increase the number of calories burned, particularly after high-fat meals.

Cinnamon C. zeylanicum Bengali Darchini constitutes the drug Cinnamon. The drug is used in diarrhoea and nausea. It is used as a stomachic and carminative, it cures gastric debility and flatulence; and also has the property of destroying certain germs and fungi.

Cinnamon was once more valuable than gold and has been associated with ancient rituals of sacrifice or pleasure. References to cinnamon are plenty throughout the Old Testament in the Bible.

Cinnamon contains from 0.5 to 1 present essential oil, the principal component of which is cinnamic aldehyde (about 60%). Other components are eugenol, eugenol acetate, and small amounts of aldehydes, ketones, alcohols, esters and terpenes. Cinnamon leaf oil is unique in that it contains eugenol as its major constituent (70-90%). The cinnamic aldehyde and/or eugenol present are both antifungal agents. Cinnamon is a stimulant, astringent and carminative, used as an antidote for diarrhoea and stomach upsets. Datura Datura Stramonium Bengali Dhatura drug consistts of dry leaves, flowering tops and seeds of the plant. The chief active principle in the leaves is hyoscyamine; the drug is, therefore, useful in the same manner as Belladonna or Hyoscyamus. The drug is useful in bronchitis or asthma, and controls salivation in mouth; it is antispasmodic and narcotic. The seeds also contain hyoscyamine and similar properties as the leaves.

Emblica Emblic Myrobalan Bengali Amloki hindiAmla fresh or dried fruits of this tree are used as laxative and in treatment of enlarged liver, piles, stomach complain, pain in eyes etc. It is a very rich source of vitamin C. Certain experiments on patients of pulmonory tuberculosis showed that vitamin C of Emblica fruits is more quickly assimilated in human system than synthetic vitamin C. Flowers, roots and bark of the tree are also medicinal, seeds are reported to cure asthma and stomach disorder (S. K. Jain, 2001).

This fruit is a great asset for the arsenic patients. Instead of takingexpensive imported tablets, Embelic is very cheap and more effective and every one can grow the plant at home.

One teaspoon of Amla juice mixed with a cup of bitter gourd juice has been recommended by naturopaths for its properties of stimulating the Pancreas which secretes insulin for reducing blood sugar. Amla seeds or dried amla is equally invaluable for control of Diabetes.

It is also effective in the treatment of amlapitta (peptic ulcer), as well as in non-ulcer dyspepsia. The alcoholic extract (1gm/kg) given to isoprotenol-pretreated rats resulted in an increase in cardiac glycogen and a decrease in serum LDH, suggesting a cardioprotective action. It also demonstrated a statistically significant reduction in serum cholesterol levels.

Traditional Uses:The fruit is commonly used in the treatment of burning sensation anywhere in the body, anorexia, constipation, urinary discharges, inflammatory bowels, cough, hemorrhoids, fever, thirst, and toxicity of the blood.

Amla is very rich in Vitamin C, It has 20 times the Vitamin C content of grapefruit and 15 times that of lemon. In dried amla (pieces or powdered), vitamins are retained and protected due to the natural antioxidant properties of the fruit Other vital benefits include:

- Cooling effect and reducing of body heat.
- Inhibiting phlegm and bile.
- Increase in the production of semen and help in urinary and gynecological problems.
- Good for pulmonary ailments Reducing of body fat Improving hair texture and eye health.

SPECIES WITH POSSIBLE DEVELOPMENT POTENTIAL: 8. MA-KHAAM POM (PHYLLANTHUS EMBLICA LINN)

Costus (English), Saussurea lappaKushtha (India):An erect robust perennial herb, the dried roots of which constitute the drug.

The roots contain resinoids, essential oil, alkaloid, inulin, a fixed oil and other minor constituents like tannins and sugars. The essential oil of the roots has strong antiseptic, disinfectant and anti-inflammatory properties. An alcoholic extract of the herb has been found very useful in the treatment of bronchial asthma.

Fenugreek (Trigonella foenum-graecum)

Fenugreek is a slender annual herb of the pea family (Fabaceae). Its dried seeds, used as a food, a flavouring, and a medicine.

The herb is a characteristic ingredient in some curries and chutneys and the fenugreek extract is used to make imitation maple syrup. Because of its high nutritive contents, it is an important ingredient in vegetable and dhal dishes eaten in India. In India, young fenugreek plants are used as a pot herb. The leaves are widely used, fresh or dried, in Indian cooking and are often combined with vegetables. Fenugreek seeds are used in a wide range of home-made or commercial curry powders.

Fenugreek was used in Middle Ages to cure baldness. It is still used in Indonesia as hair tonic. It is traditionally used to stimulate the metabolism and there by to, control the blood sugar levels of diabetic patients. It is useful in lowering the blood pressure and because of its high iron content it is also given in cases of anaemia. Fenugreek is used medicinally as a digestive aid and to promote lactation in both women and in cows. The seeds have been used as an internal emollient for inflammation of the digestive tract and as an external poultice for boils and abscesses; but their present medical use is principally confined to the treatment of cows and horses. It contains diosgenin, a compound used as a starting material for sex hormones in the pharmaceutical industry.

Ginger (Zingiber officinale) is described as acrid, heating, carminative, rubefacient and useful in dyspepsia, affilations of throat, head and chest, haemorrhoids, rheumatism, urticaria and many other diseases.. Ginger has been used as a medicine in India from Vedic period and is called great medicine. Ancient physicians used it as a carminative or anti flatulent. Galen, the Greek physician, used ginger to treat paralysis caused by phlegmatic imbalance in the body. Aviceena the Arab physician used it as an aphrodisiac centuries ago pomose also used ginger in the treatment of gout. Ginger is pungent and a bit bitter in taste. It acts as digestive, carminative, stomach, anti pyreutic, generates heat expels flatus and cough, purifies blood and is invigorating. One teaspoon of ginger juice mixed with a cup of fenugreek decoction and honey to taste makes an excellent diaphoretic mixture to proliferate sweating and reduces fever in influenza. It acts as an expectorant in bronchitis, Asthma,and whooping cough.

Gurmar(Gymnema sylvestre): An Indian folk favourite for treating diabetes. Tea made from this herb helps to boost insulin production. Research studies have also suggested that this herb may actually increase the number of beta cells in the pancreas. More studies are pending.

Garlic: Similar to onions in its action: It is suggested that raw is best or lightly cooked in food.Garlic is said to stabilize blood sugar, enhance immunity and improve circulation. (helps regulate blood sugar levels and so can be helpful in late onset Diabetes.) Garlic lowers blood-cholesterol levels-reduces hypertension-stimulates the digestive system. Garlic enhances the body's immune defenses.

Henna (Lythraceae) BengaliMehndi The trade name is based on the word Hina which is the arabic name of the drug.

The chief use of Henna is a pleasant orange dye for coloring. The leaves of the plant have certain medical properties. They are astringent and used as a prophylatic against skin diseases. The leaves have also been shown to have some action against tubercular and other bacteria, and in typhoid and haemorrhagia (S. K. Jain, 2001).

It is difficult to pinpoint the exact origin of mehendi. Some historical evidence suggests that it was introduced in India during the 12th century AD. Again, there is proof that henna was used to stain the fingers and toes of the Pharaohs prior to mummification over 5000 years ago when it was also used as cosmetic and for its healing power.

Mehendi: Tradition mingled with a new look Colour your palms with rusty-red

Hing (Ferula foetida) or Asafoetida Bengali Hingis a sedative, expectorant and laxative It is very useful remedy for relieving spasms and ingestion. flattulent colic, cholera and whooping cough. It is a stimulant for respiratory and nervous system and very effective in pneumonia and bronchitis in children.

Medicinal Properties:

- Diabetes: Mix 1/4 tsp hing powder in 2 tsp bitter gourd juice.Take twice a day.
- Indigestion: Mix 1/4 tsp hing powder with a ripe banana and eat.
- Kidney-Problems: Mix 1/4 tsp hing in 2 tsp fresh ginger juice.Add a pinch of salt and sip.

Jambol Eugenia jambolana Bengali Jam, Kala Jam the bark, fruits and seeds of the tree are medicinal. The bark is very astringent and is used in sore throats, bronchitis, asthma, ulcers and dysentery; it is also given for purifying blood, and as a gargle. The seeds are very useful in diabetics. The drug showed effect only when administered through injections not through oral administration. The anti-diabetic activity of this drug is more marked than that of Bijasal,Petrocarpus Marsupium Roxb (S. K. Jain, 2001).

Jaiphal (Myristica fragrans) is very well known in the Indian subcontinent for various medical properties. It is the most valuable medicine in dyspeptic complaints.-Prescribed in the low stage of fever, in consumptive complaints and asthma it is also used for stimulating digestion, healing choleraic diarrhoea, obstructions of liver and spleen etc.

Onions: Asian researchers fed subjects onion juice and whole onions and found that the greater the dose, the more blood sugar was depressed. (rawor boiled no difference). The active ingredients isolated are allyl proply disulfide and allicin. It is believed their action is in stimulating more insulin production.

Sandalwood Tree Chandana (India): A small to medium-sized, evergreen semi-parasitic tree, with slender branches, valued for its heartwood.

Both the wood and the oil have long been employed in medicine. The main constituent of sandalwood oil is santalol. It is used to alleviate itching and inflammation. It is credited with cooling, diaphoretic, diuretic and expectorant properties, and sandalwood finds several applications in household remedies.

Soapnut-tree (English) Sapindus mukorossi, Sapindus emarginatus, Sapindus trifoliatus, Bara rita, Ritha (Bengali), Phenila, Arishta (India): A deciduous tree found wild in north India, usually with 5-10 pairs of leaves, solitary with large drupes. This tree belongs to the main plant order Sapindaceae and family Sapindeae. The species is widely grown in upper reaches of the Indo-Gangetic plains, Shivaliks and sub-Himalayan tracts at altitudes from 200m to 1500m. Also known as soap-nut tree, it is one of the most important trees of tropical and sub-tropical regions of Asia. It is also called doadni, doda and dodan in Indian dialects.

This tree flourishes in deep clayey loam soil and does best in areas experiencing nearly 150 to 200 cm of annual rainfall. The trunk of Ritha is straight and cylindrical, nearly 4 to 5 m in height. The canopy comprising side branches and foliage constitutes an umbrella-like hemispherical top measuring about 5 m in diameter. The tree can reach an height of 25 m and a girth of 3 to 5 m in nearly 70 years of its existence. Ritha is thus an excellent tree for planting along boulevards.

Ritha flowers during summer. The flowers are small and greenish white, polygamous and mostly bisexual in terminal thyrses or compound cymose panicles. These are sub-sessile; numerous in number and at times occur in lose panicles at the end of branches. The fruit appears in July-August and ripens by November-December. These are solitary globose, round nuts 2 to 2.5 cm diameter, fleshy, saponaceous and yellowish brown in color. The seed is enclosed in a black, smooth and hard globose endocarp. The fruit is collected during winter months for seed and or sale in the market as soap nut. The trunk of Ritha is straight and cylindrical, nearly 4 to 5 m in height. The canopy comprising side branches and foliage constitutes an umbrella-like hemispherical top measuring about 5 m in diameter. The tree can reach an height of 25 m and a girth of 3 to 5 m in nearly 70 years of its existence. Ritha is thus an excellent tree for planting along boulevards. Ritha wood is hard and light yellow in color. It is close-grained and compact weighing about 30 kg per cubic foot. The wood is utilized for rural building construction, oil and sugar presses, agricultural implements, etc.

Ritha seed germinates easily. To ensure cent per cent germination, the seed is soaked in lukewarm water for 24 hours and then sown, either directly in already prepared 60 x 60 cm pits at 5m x 5m spacing or sown in polythene bags filled with clayey loam soil mixed with farmyard manure or similarly prepared nursery beds. For thousands of years Indians have been using it for a variety of purposes. It is known in Hindi as Ritha, reetha, aritha, dodan, kanma and thali. Had soapnut arrived in Britain at the same time as tea, this country would have remained far less polluted, with greater reserves of fossil fuels for the millennium ahead.

Chinese peasants traditionally used the small yellow fruit of the soap berry tree (Sapindus mukorossi) to make soap. Very easy to make: simply grinding up the rind and soaking it in water produces a soft liquid soap used for washing and as a shampoo--popular with village women because it "beautifies the skin and removes freckles".

Soapnuts have been around for a very long time in India and Nepal. People there have always been washing their clothes with soapnuts. The secret of the soapnut is as simple as it is effective: The nut shell contains saponin, which acts like soap as soon as it gets in contact with water.In fact the skin of the fruit is highly valued by the rural folks as a natural produced shampoo for washing their hair. They also use these for washing woolen clothes. This is why some botanists have named the species as Sapindus detergens.

Soapnuts have long been used in the Western world for soap production, usually together with many chemical additives which are not really necessary for the actual washing process and which are damaging to the user as well as our environment.

The percentages of individual acids were found to be: palmitic, 4.0; stearic, 0.2; arachidic, 4.4; oleic 62.8; linoleic, 4.6; linolenic, 1.6; and eicosenoic, 22.4. The oil is composed of 0.1, 2.1, 22.0, and 75.8% trisaturated, monounsaturated disaturatd, diunsaturated monosaturated, and triunsaturated glycerides, respectively. The special characteristic of the Sapindus mukorossi seed oil is its content of 26.3 and 26.7% triolein and eicoseno-di-oleins, respectively (Lipids. 1975 Jan;10(1):33-40).

Method: Soapnuts contain saponin, which works similar to soap. Ironically, soapnuts are generally used in the West to extract the saponin in order to manufacture industrial soap, whereby the original potential of its use as a laundry detergent was ignored for a long time. Once these soapnuts get in contact with water in the washing machine, the saponin is naturally extracted and creates the same effect as a conventional laundry detergent. The effect is positive: soapnuts clean remarkably well! All common stains will be removed, just as with the use of normal washing powder. Merely persistent stains, such as blood, or red wine, are more difficult to remove.

100 grams of soapnuts produces a good 2.5 litres of soapnut Juice. 3-4 spoons of Juice are sufficient for a laundry load, a little less, with added vinegar will clean a load of dishes in the dishwasher. An infusion made from soapnuts gives a shampoo which works well and effective to fight dandruff as well as gives hair a silky shimmer and vitality. After the hairwash with a soapnut infusion it is easy to comb through the hair, and it takes much longer to become oily. Very suitable especially for allergy sufferers. Soapnut also discourages the occurrence of parasites, such as nits or lice.

Uses:

- soapnut is excellent for washing and bathing humans and pets. It leaves the skin with a soft, smooth layer which protects against infections and insects.
- mechanic's hands, stained hands, or those where the skin is cracked from chemical cleaners can gain considerable relief. noticeable improvements within two weeks have been found, including smoother skin and the removal of ingrained marks.

- soapnut is a natural exfoliant. It is considered to be second to none and is also very common in the Indian Ayuverdic healing system.
- in hair care, soapnut helps to remove dandruff, gives hair more body and works against infections of lice and other parasites. It leaves the hair, not just looking healthy but, actually healthy. Recently there has been evidence showing that soapnut also reduces hair loss.
- soapnut is traditionally used as a natural and effective treatment for skin complaints including eczema, chronic itching and psoriasis.
- soapnut is perfect for washing clothes, with no optical whiteners, foaming agents or other chemical additives. In Nepal, soapnut is used for washing the finest silks and woollens in preference to any other product.
- elsewhere in the kitchen, soapnut is also invaluable; dishes, cutlery and even greasy pans can be cleaned with soapnut-and it is dishwasher friendly.
- most of us are unaware that many of the fruit and vegetables we eat are grown using quantities of harmful chemicals. supermarkets also use chemicals to increase their shelf-life, hence their recommendation to was fresh produce before use. Scientific test have shown that a ten minute soak in a soapnut solution will remove up to 95% of the surface pesticides and chemical residues.
- other uses include cleaning teeth, polishing jewellery, cleaning glass, paintwork and even washing the car!
- in the garden a soapnut solution can be used as a spray to repel and prevent a wide variety of pests and blight, including aphids and blackfly. A well regarded scientific horticulturist is currently researching these claims, with great success.

Especially for allergic persons, persons suffering from neurodermatitis and people with sensitive skin, chemical detergents often provoke an aggravation of their ailment. Furthermore, it is evident that some of the chemicals used in some detergents are allergic. In our civilized surrounding, the amount of allergic substances rises steadily.

The fruit is valued for the saponins (10.1 %) present in the pericarp which constitutes up to 56.5 per cent of the drupe. The fruits are credited with expectorant and emetic properties and are used in the treatment of excessive salivation, epilepsy and chlorosis.

The powdered seeds are said to possess insecticide properties. They are employed in the treatment of dental caries. It cleanses the skin of oily secretion and is even used as a cleanser for washing hair and a hair tonic, and forms a rich, natural lather.

Its detergent action, which cleanses the hair and removes, accumulated debris and a sebaceous material further more imparting speculiar reflection and hair luster.

Squill Urginea Indica Bengali Ban Piag is used in ailments of heart, and in cough and bronchitis. It promotes urination. Clinical trials have confirmed efficacy Indian Squill in chronic bronchitis and bronchial catarrh.

Tamarind (Tamarindus indica), Bengali Tetul is one of the most beautiful trees of the Indian subcontinent. An ancient Sanskrit script describes the fruit as 'refrigerant, digestive, carminative and laxative' and useful in bile-related diseases.

Tamarind is semi-evergreen, tropical tree that grows to about 24 m (80 feet) tall and has long drooping branches with alternate, pinnately compound (feather-formed) leaves; the leaflets are about 2 cm (0.75 inch) long. The yellow flowers, about 2.5 cm across, with a red stripe are borne in small clusters. The dark brown fruit is a plump pod 7.5-24 cm long that does not split open. It contains 1 to 12 large, flat seeds embedded in a soft, brownish pulp. This pulp has a high tartaric acid content, that imparts for its sourness.

Tamarind is a good laxative and an antiseptic. It is used for tummy upsets and for the treatment of ulcers. Over-ripe fruits can be used to clean copper and brass.

Foods: Such as broccoli, nuts, oysters, mushrooms, whole grains, wheat cereals, rhubarb and brewer's yeast. Broccoli is particularly rich in chromium, as is barley. What it is: A MINERAL that makes the body more sensitive to insulin, regulates CHOLESTEROL and fatty acid production in the liver, and aids in the digestion of PROTEIN. If chromium is lacking, blood levels of cholesterol and fatty acids rise, GLUCOSE is poorly metabolized, and, in severe deficiencies, there may be nerve damage.

Spices and herbs used in Indian Sub-continent:

Name	Botanical Name
Amchur	Mangifera indica
Anadhana (Anardana)	Punica granatum
Anise (sanof)	Pimpinella anisum
Asafoetida (Hing)	Ferula foetida
Bay Leaf (Tejpata)	Laurus nobilis
Black cumin (Kala jeera)	Carium nigrum
Capsicum (chilli, cayenne pepper etc))	Capasicum frtescens/annum
Caraway (Shah jeera)	Carum carui
Cardamom (Elachi)	Amomum subulatum
Cardomom Brown (Bara Elachi)	Amomum subultum
Carum (Ajwain)	Carum copticum
Chirauli Nut (Charoli)	Cinnamomum zeylanicum
Cloves (Laung	Caryophyllus aromaticus
Coriander Dhania)	Coriandrum sativum
Cumin (sada jeera)	Cuminum cyminum
Curry leaf (Kurry Pata, mitha neem)	Murraya koenigii
Fennel (Sonf)	Foeniculum vulgare
Fenugreek (Methi)	Trigonella foenumgroecum
Garlic (rashun)	Allium sativum
Ginger (Ada/Adrak)	Zingiber offcinale
Mace (jaiviri)	Myristica fragrans
Mint (Podina)	Mentha viridis
Mustard Brown (Rye)	Brassica juncea
Mustard Black (Sarson)	Brassica nigra
Nigella (Kalo jeera), wild onion seed	Nigella saitva
Nutmeg (Jayphal	Myristica frarans

Onion (Piyaz)	Allium cepa
Poppy Seeds (Khas Khas)	Papaver somniferum
Ratanjot	Onosma echioides
Safron	Crocus sativus
Sesame (Til)	Sesamum indicum
Tmarind (Tetul, Amli)	Tmarindus indica
Tejpat (Tejpata)	Cinnamomum tamala
Turmeric (Haldi)	Curcuma longa
Sacred Basil (Tulsi Pata)	Ocimum sanctum
Sweet Basil (Kala tulsi)	Ocimum baslicum
Dill (Sowa)	Peucedanum graveolens
Mint (Pudina)	Mentha arvensis

PLANTS-THE GREATEST BIO-CHEMIST OF THIS PLANET

Plants are wonderful chemists, a trait that benefits not only the plants themselves but also human. Although the chemicals in plants are sometimes deadly to the animal, the same product in the leaves, roots, bark and flowers can be useful medicines for people when the doses are designed for the human body. Only less than five per cent of the plant kigdom have so far been analysed as potential medicine and remaining 95 per cent are still to be analysed.

The Heart Regulator

Millions of heart patients treat their heart ailments daily with medical derived from the delicate foxglove plant, Digitalis. The secret of this plant's success in treating heart disease lies in the glycosides found within its leaves. Of millions of patients who have been prescribed Digitalis, about 20 per cent experience side effects. Since we already know of other species that contain cardiac glycosides that would cause fewer side effects.

Cinchona against Malaria

The Peruvian natives called the tree quina-quina-from which the name quinine was drawn. After the initial discovery of quinine, chemists isolated 40 different alkaloids from Cinchona. One additional alkaloid, quinidine, successfully treats certain heart disease. During World War II synthetic product replaced the highly affective natural quinine.

Synthetic Replacement-Not Perfect Substitute

In tropical countries malaria can no longer be contained with synthetic quinine. Natural quinine, however, is still highly effective against these strains. As we see with Cinchona, synthetic replacements are not perfect substitutes for nature. What happened with this anti-malarial drug could happen with any synthetic medicine.

Alpha-terphinel-Fight Against Malaria

During the 1970s many countries suffered a huge increase in the number of malaria cases and for some this amounted to up to 30 times the low figures achieved during the 1960s. In India today more than half the nation's health budget is spent on anti-malaria campaigns. Its record of cases over the last three decades makes alarming reading; from 100 million in 1952, down 60,000 in 1962 and up again in 1978 to 50 million. Alpha-terphienyl, a compound derived from Tagetes marigolds, seems to be as effective as DDT in controlling mosquito larve.

Jambul Fruit-Eugenia Jambolana Reduces Suger

The jambul tree is found mainly in India and the fruit and its seeds have been revered for centuries for their medicinal properties. This fruit reduces the sugar in the blood and is very good in the control of diabetes. Its seeds contain Glucoside, Jamboline and Ellagic acid which are reputed to have the ability to check the conversion of starch into sugar in case of excess production of glucose. Therefore, Jambul seeds are also used as a remedy for Diabetes. Use of Jambul seeds for Diabetes was also confirmed by "Shaligram Nighantu Pharmacopia" in ancient India. Jambul seeds reduce urine sugar quickly and have been very effective in controlling diarrhoea. Other constituents of the fruit include Resin, albumen, gallic acid, essential oil and tannic acids.

Green Tea

Studies confirm that tea catechins-potent antioxidants-are effective in suppressing increases of glucose and insulin concentrations in the blood. Since blood sugar tends to increase with age, this effect is an extremely important anti-aging benefit

(Horigome, T., Kumar, R and Okamoto, K.: Brit J. Nutr., 60,275-285,1988)

Plus, tea polyphenols inhibit the activity of amylase, a starch-digesting enzyme found in saliva and in the intestines. Starch is broken down more slowly, and the rise in serum glucose is minimized, so that you don't crave sweets and other snack foods after eating a meal.Since insulin is our most fattening hormone and, with cortisol, our most pro-aging hormone, if you drink Green Tea or take its extract in the form of a nutritional supplement, you gain a wide range of benefits that accompany calorie and insulin control. This "starch blocking" effect of green tea may be part of the reason Japanese people living in Japan can eat so much rice but remain thin. They have a tradition of drinking green tea before every meal. The antioxidants in Green Tea also help reduce the oxidation of low-density lipoprotein (LDL) or "bad" cholesterol, a process that can lead to clogged arteries (Luo, M., et al. "Inhibition of LDL oxidation by green tea extract." The Lancet 199, 349:360-361).

Traditional Heritage for Today and Tomorrow

Plants are potent biochemists, man is able to obtain from them a wondrous assortment of industrial chemicals. Plants produce these chemicals as they absorb energy from the sun and convert variety of substances that appear to be infinite. Less than five per cent of all plant species have been analysed as potential medicine.

Potential new plant drugs drawn from the 95 per cent of the plants still to be analysed. As the world's human population continues to explode conditions for the wildfire spread of new microbial disasters also increases. We should seek out plants that are most likely candidates to combat the predicted disease of the future. A modern example is the current AIDS epidemic that is sweeping the world. Scientists are looking for natural compounds with right antiviral properties. One of these is castonospermum ausrale, a black bean tree grows in rain forest.

The high species diversity on coral reef gives rise to another, often over-looked, benefit: their potential as sources for new drugs.... Researchers have discovered that number of these highly active compounds may have useful medical application.

Vegetables Consumption Lowest in Bangladesh

The amount of consumption of vegetable is lowest in Bangladesh among South and South-East Asian countries. It is only 28 gms. per head per day. Whereas in China it is 500 gms. and in South Korea it is 600 gms. The daily requirement of vegetables for a person is 250 gms. It was stated by the speakers in the concluding session of a week-long training course on 'vegetables seed production' held at Bangladesh Agriculture Research Institute (BARI), here on May 27, 2004. In Bangladesh the rate of use of better quality seeds is also lowest in South and South-East Asian countries (Daily Observer,June 6, 2004).

1. Bangladeshi Jute (Corchorus capsularis, Corchorus olitorius) Leaf as Medicine
2. NEWS ON HERBAL MEDICINE
3. Tribal cures for modern ailments.

WATER HYACINTH EICHHORNIA CRASSPIES A NEW RESOURCE

Since early nineteenth century exploreres charmed by the enchanting beauty of waterhyacinth flowers in the vast luxirant rainforest of Amazonia Basin and Central America were preserved in all botanical gardens of Europe. At the end of nineteenth century it came to open water body in the Indian Subcontinent that was imported by the ruling British Authority. In a very short time the plant spread out so vigorously that caused serious problems in utelization of water resources

Since its invasion in the tropical and subtropical countries several billions dollars were spent to destroy it but any success was noticed. In Bangladesh waterhyacinth creates fouling of drinking water by their decomposition, obstruction to flow of water and navigation in rivers, competition with agricultural products, fish kills due to depletion of oxygen in water, and spread of epidemics by supporting variety of harmful animals including several disease vectors

Recent studies (Dinges 1981; Gilman et al., 1981; Forgione et al., 1982; Parashar, 1970; and Trivedy et al., 1985) suggest that waterhyacinth can control water pollution caused by the disposal of sewage and other wastewater effluents from different sources, domestic or industrial and also by the runoff from agricultural fields.

The Wastewater Engineering faculty of the University of Florida set up the first large scale waterhyacinth waste water system in Coral Spring. The system comprised of five asphalt-lined 38 cm deep ponds with a total area of 0.5 ha. with daily inflow of 387.5 m and 6 day retention period. Sweet (1979) reports that the system removes 93% of influent BOD, 67% of TSS (total suspended solids), 97% of total nitrogen and 79% of total phosphorous. About 15-20% of the plants were harvested monthly and used for making compost.

Most studies suggest that a simple passage of wastewater through a waterhycinth pond improves water quality. The mechanisms involve in wastewater purification using waterhyacinth are similar to conventional treatment facilities.

Fertilizer

Water hyacinth cmpost fertilizer contains:

- 30% Potash;
- 7% Phosphoric Acid;
- 13% Lime (IUCN, 1992).

Its value has been proved in Sudan, where it has increased peanut production by over 30% (Maltby, 1986). In developing countries the weed can be used for waste water treatment facilities and untreated weeds can be a gold mine for producing bio-fertilizer. The industries of industrial countries are interested to transfer expensive and ever dependable technology to the third world countries. Water hyacinth is cheap and affordable simple methos for developing countries.

Carcasses (cicada carcasses) Return Nutrients to Soil

Americans from Maryland to Indiana will have to fend off clouds of cicadas, insects with transparent wings, black bodies and red eyes which dig themselves out of the ground every 17 years to mate before dying. The insects, which make a deafening buzzing sound as they reproduce, have begun to emerge in massive numbers. Scientists have assured the deeply bug-averse that the cicadas are harmless insects solely interested in mating and laying their eggs for the next three weeks only to vanish again until 2021.

Their decaying carcasses gave a super-size boost in nutrients to forest soil and stimulated seed and nitrogen production in a plant important to the forest ecosystem, researchers reported in issue of the journal Science. The findings might explain why tree growth increases for several years after a major cicada emergence, experts said. Bard College professor Felicia Keesing likened it to someone pouring a pound of fertiliser per square yard over the forest floor. She co-wrote an article accompanying the research paper on the impact of cicada carcasses on soil and plants. Yang experimented with naturally occurring densities of cicada carcasses of as much as 300 bugs per square yard. For each density he measured the soil's nitrogen and bacterial and fungal growth over varying periods of time after carcasses were applied. Soil content of a form of nitrogen used by plants was many times higher-199 per cent to 412 per cent-in ground littered with cicada carcasses. Bacterial and fungal growth also increased. He added 140 cicadas per square metre to a plot that contained a forest plant called the American bellflower (Campanulastrum americanum); the plants later had 12% more nitrogen in their leaves, compared with plots without cicadas, and produced seeds that were 9% larger Yang's findings that decaying cicada carcasses apparently stimulate a rush of soil nutrients might explain why other analyses have shown that tree-ring growth among oaks in areas infested with 13-year and 17-year cicadas increased for the first four years after cicada emergence, Keesing said. Cicadas spend most of their time underground as nymphs, sucking on tree roots and diverting some of the nitrogen that would otherwise go to the plant. In their last few months they emerge from the ground, crawl up trees and shed their hard skins. Over the next few weeks they sing to attract a mate (The Associated Press, December 1, 2004).

CONSERVATION OF MICROBIAL BIO-DIVERSITY

Home garden diverse plantation will not only conserve threatened plant species, it will also attribute to increase diverse microbial bio-diversity that seldom mentioned or no attention in overall reviews of biological diversity. In ecosystems, micro-organisms are are important as symbionts (endophytes, mycorrhizae, and in insect guts), in nitrogen fixation (rhizobia, cyanobacteria, cyanobacteria-containing lichens), in the biodegradation of dead animal and plant material, and in controlling the size of populations of plants and insects through natural bio-control. The applications of micro-organisms in the bio-control of pests and weeds are becoming increasingly recognised. In developing countries, major short-and medium-term benefits can be expected from improve inocula for mycorrhizae and nitrogen-fixing Rhizobium strains; these improve tolerance to environmental stress and reduce the need to apply artificial fertilisers, repectively (Mantell, 1989).

BIOTECHNOLOGY

In1919, Karl Ereky coined the term "biotechnology" to refer to the interaction of biology with human technology. Thus one definition of biotechnology is that it is a method through which life forms (organisms) can be manipulated to provide desirable products. Crops have been the subjects of modification and adaptation to human needs since the beginning. The development of the science of genetics in the 20th century was an important factor in plant breeding programs that have produced the remarkable diversity of fruits, vegetables and grains we enjoy today. Through the manipulation of genomes, chromosomes or single genes, plants can be adapted quite precisely to specific purposes. Today, a discussion about the future of agriculture cannot be divorced from modern biotechnology. Supporters of GM food say, it could reduce the use of pesticides and fertilizers, allow people to farm in harsh environments and increase crop yields. It could also make our food healthier and more nutritious to eat:

Protein factories

Plants and fruit could be turned into biological factories to harvest custom proteins and materials such as:

- Vaccines and drugs
- Healthy oils to combat illnesses such as heart disease
- Eco-friendly biofuels

- Bio-lubricants to replace current hydraulic fluids
- Biodegradeable plastics

Opponents see it differently-they say that no one can predict the long term impact of GMOs on other plant life and on the health of the soil. There's some evidence of irrevocable soil bacteria adaptation already. They also object to the patents and licensing agreements which, they maintain, deprive farmers of control over their livelihoods. Traditional plant breeding involves crossing of different plants with useful characteristics, and has been very Although still a relatively young scientific discipline, genetic engineering is being applied in a broad variety of ways, mainly in the biomedical research, but also in agriculture and food production. With the help of genetic engineering it is now possible to transfer for example, a gene from a bacteria to a plant so that the plant produces the corresponding bacterial protein in its cells. A plant genetically modified in this way is also called transgenic plant.

GM Crops in Bangladesh, Looming Danger

Land-strapped Bangladesh is set to grow genetically modified (GM) crops to augment food production to meet the growing demand of a growing population. To start with, four types of crops would be developed soon by applying biotechnology under the National Agriculture Research System (NARS). These are drought-and saline-tolerant rice, late blight resistant potato, fruit and shoot borer resistant eggplant and pod borer resistant chickpea. Steps will soon be taken to set up 'containment facilities'--a safety infrastructure to prevent any jumping of modified genes to nature--at particular NARS institutes such as Bangladesh Rice Research Institute and Bangladesh Agriculture Research Institute. It will take at least two years for commercialisation of the seeds developed at these institutes.

Agricultural Biotechnology Support Project II (ABSPII), funded by the United States Agency for International Development (USAID), will support the endeavour. Cornell University of the USA is managing the project. A donor-aided workshop on agricultural biotechnology yesterday revealed the plan under which genetic modification would be carried out to gain special disease-fighting traits in crops FBCCI chief Abdul Awal Mintoo, who also heads East-West Seed company, said the advent of agricultural biotechnology would facilitate further genetic improvements of seeds (Daily Star, October 7, 2004).

Half a century's Green Revolution in Bangladesh and other countries, and the current biotech moves-hybrid seeds plus costly agro-chemicals and irrigation, which apparently yield handsome harvests-is now known to hurt the seed base and the soil fertility in alarming ways. Besides they make crops more pest prone and thus more dependent on pesticides.

The whole history of agriculture is one of experiments. In ancient times perhaps it was a trial-error process, today it is a more organised and documented process. We do not know how many of those experiments failed. It is those which were successful that have trickled down to us. Yet in this Third Millennium because of a number of failures, we tend to fear and fight against any innovative method. Bovine growth hormone (BGH) produced through biotechnology can increase the milk yields of cows by 5 to 20 per cent. But BGH use has also resulted in the increased incidence of mastitis in cows (Afroza Quadri, November 10, 2004).

1. Potential Hazards from Transgenic Crops
2. Farmer Liability and GM Contamination-Schmeiser Judgment.

Greater yields, bigger profits, easier farming methods.. In India, where there are nearly a billion mouths to feed and two-thirds of the population is involved in farming, the promises of the multinational seed companies are enticing. Yet the so-called wondercrops could destroy rather than improve the livelihoods of India's small farmers.

Farmers to Lose Control over Seeds with use of GMO'

Many farm labourers might lose their jobs and bio-diversity would be lost with the massive usage of genetically modified organism (GMO) in the country, said the speakers at a workshop yesterday. They said some multinational companies are trying to control world's agriculture and their emergence here will drive out the individual farming. The speakers said patent rights will come straightway with the introduction of GM seed and farmers will lose the authority over the seeds. They said only 11 companies are now controlling the entire seed market all over the world. The workshop titled 'Agriculture, Food and Commerce: Peoples' Movement' was organised jointly by Ubinig, Coastal Development Partnership, Lokoj and Uttaron in association with the International Food Security Network and Action Aid in the city. The speakers strongly opposed the world trade organisation (WTO) agreement on agriculture.

Terming the agreement partial they said fish and jute, two main agricultural products of Bangladesh, are not included in WTO agreement. The speakers demanded immediate cancellation of the agreement that signed by government in 1995. They also demanded ban on imported GMO and hybrid agricultural product, and urged all to take stand against the introduction of golden rice in the country.

"On one hand the donors keep telling our government to cut subsidy on agriculture but on the other hand they are subsidising their agriculture sector heavily to get our market, the speakers alleged. They said they will make a proposal based on the opinions of mass people before Hong Kong WTO conference (Daily Star, August 01, 2005).

Monsanto's latest flagship technology makes a nonsense of its claim that it seeks to feed the worlds hungry. On the contrary, it threatens to undermine the very basis of traditional agriculture-that of saying seeds from year to year. What's more, this "gene cocktail" will increase the risk that new toxins and allergens will make their way into the food chain. The Terminator does more than ensure that farmers can't successfully replant their harvested seed. It is the "platform" upon which companies can load their proprietary genetic traits-patented genes for herbicide-tolerance or insect-resistance-and get the farmers hooked on their seeds and caught in the chemical treadmill (The Ecologist, USA, Sept-Oct 1998 v28 n5 p276(4)).

GM Crops Lead to Herbicide-Resistant "Superweed" in UK

British agricultural scientists have found that a genetically modified (GM) variant of rapeseed has cross-fertilized with local wild charlock plants, creating a herbicide-resistant "superweed" in the process. The transformation of a plain charlock into a superweed is something scientists had thought to be "virtually impossible." (Our Planet, August 2005)

INVASION OF NATURAL ECOSYSTEMS BY PLANTS

A few months ago an educated very wealthy person was very proud to present his garden in Savar, Dhaka, "I have got all types of plants, flowers from all over the world!" In Bangladesh influential persons can bring any type of plants through the customs without quarantine. Illegal import of plants can bring one of the worst disasters in Bangladesh.

Imported Trees

Bangladesh with the advise of the western experts planted many different types of imported trees (financed by the World Bank, Asian Development Bank etc.)without studying impact on ecology.In the past few years the banana, pineapple and papaya cultivators have illegally cleared thousands of acres of forestland. The clearing continues unabated. Given the trends, demise of the Modhupur forest is imminent. This is happening with the Forest Department officials, employees and guards around

Quarantine centres under the Department of Agriculture Extension (DAE) are allegedly issuing indiscriminate import permission certificates and release orders for import consignments of all kinds of agricultural products, irrespective of whether they are of banned varieties or not. Sources in the DAE have told New Age(February 6, 2003) that some officials at the plant quarantine centres in Dhaka, Benapole, Darsana, Chittagong seaport, Bhomra and Hili, including some service centres, are issuing import certificates and release orders to such imports disregarding the obligatory examination of the plants.

It is further alleged that some officials of the DAE are issuing the phyto-sanitary certificates without any examination of the products.

The city witnesses many foreign trees which is again harmful. It is the splendid creation of nature that all the trees are not suitable for all kinds of environment. The trees of cold countries are adapted there to clean and make the environment sound of those areas. The same kind of trees may prove dangerous in the tropical and sub-tropical areas. Only because of beauty, foreign trees should not be planted. Indigenous trees must prevail and dominate the flora and fauna of our city. Because of commercial or political gain or benefit we must not take this kind of heinous decision at the cost of our lives.

Threats from the World Bank, ADB, and IMF

The biggest threats to public forests today are industrial plantation for production of raw materials for the pulp and paper mills, and commercial plantation for production of fuelwood," said Gain. "Industrial and commercial plantations are generally short-rotation. Between harvest and subsequent rotation of plantations, the land grabbers get an apt opportunity to put their hands on public forestland in collusion with the dishonest Forest Department people.

Forests are our mother stocks of species and seeds. We can plant trees, but we cannot create forest. It is very important that we try to save our last forests. Given the rapid destruction of the forest and loss of habitat it is difficult to save our last forests. We, the outsiders want to conserve the environment often with plantation. But the forest-dwelling Adivasis are part of the environment. They believe they are the custodians of the forests. Forests with the Adivasis and their traditions are most diverse, on the other hand, Plantations depict an opposite picture. These seriously lack biodiversity and cause massive soil erosion. Infrastructure, construction, and tourism are not eco-friendly. These have destroyed our forests and caused havoc for the forest dwelling Adivasis. Population increase, mainly for in-migration of Bangalees from the plains and flawed land use policies have upset the balance in the Chittagong Hill Tracts (CHT). Jumias are wrongly blamed. Plant and crop diversity have decreased in the CHT because of decrease in jum cultivation, reduced fallow period, competition for land and tendency to control jum. Not the jumias but the outsiders are behind these factors.

Among the major institutional threats to the forests are the World Bank and ADB (Asian Development Bank) (Dr. Khaled Misbahuzzaman, Chittagong University, July 16 2004). Destruction of our forests is caused largely by wrong prescriptions that come from the World Bank, ADB and IMF. A solid example is the destruction of the mangroves for shrimp cultivation in the coastal areas (Daily Star, July 16, 2004).

1. Rich and local strongmen ruthlessly trying to elbow out thousands of landless families from the chars (islands)
2. Plantations (replacing native species) Are Not Forests
3. Third World communities fight the "Blue Revolution"

We must unite, resist the misdeeds done to the forest and forest people, educate ourselves and do whatever else we can to save our last trees.

"Sissoo Trees" (Dalbergia sissoo Roxle) is an example that leading to serious concerns about the future of the country:

Sissoo (Dalbergia sissoo Roxle) Trees

At a time of growing global awareness about the environment biodegradation continues in Bangladesh leading to serious concerns about the future of the country. According to an exclusive news item published in The Independent a large number of Sissoo trees are dying in the northern districts of the country being afflicted with the fungal disease 'dieback'. This by no means is a new phenomenon with the disease killing the trees, which promised to bring a fresh economic impetus in the region, for the last eight years. Despite assurances of foreign aid regarding the matter the authorities are doing nothing to stop the onslaught of the disease.

Killer Disease Attacks Sissoo (Dalbergia sissoo Roxle) Trees

The disease has broken out in an epidemic form in 18 districts. The affected districts are: Rajshahi, Naogaon, Chapai Nawabganj, Natore, Bogra, Pabna, Sirajganj, Rangpur, Gaibandha, Dinajpur, Kushtia, Chuadanga, Meherpur, Jessore, Jhenidah, Faridpur, and Dhaka. In the dieback disease, first root infection occurs which results in the rotting of the root. Then leaf shedding is followed by the death of the affected as well as neighbouring trees progressively. Along with death of some of the branches at the early stage, a characteristic pink to reddish-pink liquid is seen oozing out of various places on the stem. According to a recent study carried out by the Village and Farm Forestry Project (VFFP), a project of the Switzerland Development Cooperation (SDC), over the last 15 years, 80 per cent of the trees planted in North Bengal were of the Sissoo

(Dalbergia sissoo Roxle) species. Owing to the dieback, farmers are felling young trees and as a result facing severe economic loss. On the other hand, they have to use the trees as fuel wood as they have no timber value.

The disease is not only causing economic loss but also having a great impact on the environment in the whole region. Sissoo is found to be the most successfully grown tree in the Barind and other areas of North Bengal and the southwestern districts. Without Sissoo, the whole area, especially the Barind, might turn into a desert, experts fear. Because of the disease, production and sale of the Sissoo seedlings have declined since 1996. Again, the economic loss caused by the devastating disease is colossal though there is no official statistics on the number of dead trees. One 15-year old Sissoo tree sells at the minimum rate of Tk 10,000 in the local market. A study carried out in 2001 by Bangladesh Forest Research Institute (BFRI) on the extent of damage caused by the dieback revealed that on an average 43 per cent of sissoo trees had died in 16 districts.

The districts and the mortality rates of the trees are: Chuadanga (63 per cent), Meherpur (57 per cent), Rangpur (57 per cent), Kushtia (55 per cent), Jessore (48 per cent), Jhenaidah (47 per cent), Rajshahi (41 per cent), Pabna (36 per cent), Bogra (28 per cent), Dinajpur (31 per cent), Dhaka (30 per cent), Magura (24 per cent), Faridpur (22 per cent), and Mymensingh (21 per cent). But, the forest department cannot say how many trees there are in the districts as "they do not have any statistics on this". A government statistics provided by the VFFP said that since 1985, about 2,00,00,000 trees have been planted in the Barind region-Rajshahi, Naogaon and Chapai Nawabgnaj districts-by the Barind Multipurpose Development Authority alone of which more than 80 per cent was Sissoo.

Besides, trees are also produced and planted under private initiatives. Another report of VFFP says that about 1800 private nurseries are supported by it, and they produce 50 million saplings of wood trees and fruit trees per year which are sold in the market.The problem is not unique in Bangladesh. It is also reported in India, Pakistan and Nepal.

Research Survey carried out a study on the dieback of Sissoo which revealed that different types of pathogens including

Fusarium solanii (most common), Phytophthora spp and Ganoderma were directly and indirectly involved with the disease. "The dieback disease might have link with water stagnation. Besides, Boron deficiency can also cause the harm", Farid Uddin Ahmed, a forester cum environmentalist who works with Bangladesh Agricultural Research Council and VFFP, told The Independent.

Though insects were found associated with the disease, it was concluded that insect infestation takes place as secondary infection. However, they have some role in the development of fungus in the damaged parts of the trees and fungus grows from the exposed damaged parts of the trees. The study also inferred that adequate attention was not paid to the quality of seed, site quality, site matching spacing and other factors adding that such plantation never followed timely and appropriate management guidelines. Besides, the study suspects that monoculture of Sissoo trees must be another cause of the dieback. Since 1996, saplings of Sissoo trees of 10-15 years old have been reported to be dying from dieback disease. Many farmers have reported that such trees die in about three to four months. No other timber species, except teak, is so extensively planted in Nepal, India, Pakistan and Bangladesh as Sissoo (The Independent, 2002).

This attitude is particularly disturbing considering the fact that the Sissoo trees were planted with government patronage. The northern region, already has little tree cover compared to the rest of the country and if the trees in question have to damaged because of the ailment, it may well lead to irreversible environmental damage

Environment in Bangladesh has rarely managed to find a place in the list of priorities of the policymakers. It is easy to understand the reasons in a poor country but ignoring the environment can have disastrous impact on our future and is something we can ill afford. The impacts of environmental pollution may not be immediate but are sure to cost us heavily in the long run.

As we do not have adequate forest cover it is imperative that we plant more trees and preserve the old ones as much as possible. But though there have not been shortage of people exhorting about the beneficial aspects of trees few are ready to take effective actions to protect these trees that have over the years dwindled in numbers and Bangladesh, a country known for its green foliage is now facing the previously considered improbable threat of desertification in certain areas. Indiscriminate felling of trees for use as timber and charcoal means a quick buck for many unscrupulous traders and a bleak future for our environment. Strict laws do exist against illegal logging but like in many other fields it is the implementation of laws where things go awry. (Source: Editorial, The Independent, 20. 12. 02)

Imported Products

In the case of imported products, importers have to pay an additional amount of money for taking their permission certificate and simultaneously in getting the release order for the imported goods. Those go without tests, but nevertheless are released. Quarantine centres across the country are to oversee and test the imported products in order to rigorously ensure through bio-chemical tests that those are free from germs or diseases. Bangladesh imports all kinds of agricultural products like rice, wheat, daal, cotton, varieties of fruits, saplings without soil and all kinds of spices from different countries.

A DAE entomologist said that the quarantine centres are not properly upgraded as the government alone could not do enough to bear all the necessary costs of proper maintenance (New Age, February 6, 2004)..

Invading plant pathogens have led to some of the most serious disruptions of natural ecosystems ever recorded. For example, the devastating impact of chestnut blight on North American hard wood forests and of Phytophthora root on Western Australian jarrah forests. Chestnut blight virtually eliminated the American chestnut throughout its natural range.

In Bangladesh we name our children after beautiful trees, flowers, and nature we live in. Love to nature is rooted in our blood. The use of dangerous pesticides for conserving food, wood against skin diseases etc. in Bangladesh/India is accounting for several deaths and health hazards. Traditionally useful plants will not only bring additional income to our rural poor but will also keep us happy and healthy.

7

Biotechnology : Fungicides and Insecticides

Our doctors are surprised with the rising rates of obesity, diabetes, hypertension, heart disease and cancer. Most obese women are not aware that they are diabetics till visit the doctor for some illness or surgery. Hypertension too, being a gradual rise of blood pressure, most people are not aware until they feel dizzy, and some end up with heart attacks and strokes. Knowing about nutrition, healthy living and longevity today has become so specialized that most people tend to get confused not knowing whom or what to follow. There is a diet revolution today and the more you read about cholesterol, saturated fats, butter, margarine, carbohydrates, etc. the more you get confused. The answer is, read every bit you could gather through the Internet, magazines, newspapers, and nutritionists.

Americans have given us their heritage by the large imports of wheat flour for our traditional foods for over a century. The process that wheat goes through to obtain that fine white flour we import from the States are as shocking as the process margarine goes through to make it appear like butter. Even before they are planted in the ground, wheat seeds receive an application of fungicides and insecticides. Fungicides are used to control diseases of seeds and seedlings; insecticides are used to control insect pests, killing them as they feed on the seed.

Some of the chemicals used in commercial wheat crops are disulfoton, methyl parathion, chlorpyrifos, dimethoate, diamba and glyphosphate. These chemicals increase the toxic load in our bodies, leading to increased susceptibility to neurotoxic diseases as well as conditions like cancer. Many of these pesticides function as xenooestrogens, foreign oestrogen that can reap havoc with our hormone balance, states Jen Allbritton, a certified nutritionist in the States. He further states that the researchers speculate these oestrogen-mimicking chemicals are one of the contributing factors to boys and girls entering puberty at earlier and earlier ages. They have been linked to abnormalities and hormone related cancers including fibrocystic breast disease, breast cancer and endometriosis.

Farmers also apply hormone like substances or plant growth regulators that affect wheat characteristics, such as time of germination and strength of stalk. Cycocel is a synthetic hormone that is commonly applied to wheat.

During the process of storage of wheat, the collecting bins are sprayed with insecticide, inside and out. More chemicals are added while the bin is filled. These so called "protectants" are then added to the upper surface of the grain as well as four inches deep into the grain to protect against damage from moths and other insects entering from the top of the bin.

One could imagine how much poisonous linoleic acid gets into the system by regular filmgoers, especially the kids who consume big loads of popcorn during the intervals.

Two slices of whole-meal bread with lentil curry and pol sambol, would be an excellent morning choice. Pol sambol is nutritious, scraped coconut in addition to its nutrients, antioxidants, has lauric and capric acid having anti-microbial properties. What better way to start the morning with food that kills viruses, parasites, bacteria and fungi? The chillie powder in the sambol has its vitamin C, and increases the general metabolism, also accelerates the functions of the glands. Added lime juice gives the fruity taste and vitamin C. The Maldive fish adds the omega 3 oil, good for heart health. Foods are now rated on a scale called glycaemic index, or simply 'GI'. The carbohydrates that are high in GI are less healthy. White bread (GI of 70) is used as a reference sometimes instead of glucose. The foods that take longer to be absorbed are called 'low GI (GI less than 55). The carbohydrates that are quickly digested and absorbed are called 'high GI' (GI greater than 60).

Lentils are nutritious, and have a GI of less than 30. It's a good protein for the vegetarians and diabetics. Mun-ata (green gram), lunu miris, and scraped coconuts are excellent foods as a stomach filling breakfast. Mun-ata also has a GI less than 30. Other low GI foods are Soya beans (GI 14); Peanuts (GI 15); Red lentils (GI 18); Kidney beans (GI 27) Apples (GI 38); Spaghetti (GI 41); Orange juice (GI 46); Raw carrots (GI 49); Cooked carrots (GI85); Baked beans (GI48), and so on.

Some studies have shown that a diet based on low GI foods can lower the blood cholesterol, lower risk of heart disease. It is clear that if one craves for starchy traditional foods, home-pounded rice flour would be a better choice than the imported white wheat flour from the States.

SAVE OUR GENETIC RESOURCE

I was surprised to see at Aricha Ghat, Bangladesh that vegetable and fruit sellers were shouting, "Buy Desal (country or wild variety) of fruits!" These are more expensive than imported modified varieties. But taste better and contain more vitamin and minerals but smaller in size. Local varities iof rice or pulses, spices in village market are more expensive. But people have realize that traditional varities are better. But the big NGOs are selling imported manipulated seeds promising high yields which need chemical fertilizers and pesticides.

Farmers of Indian Sub-Contient used to plant traditional seeds but now after introducing high yield varities there is always an artifical seed crisis. Whenever there is a shortage of a commodity, black marketers lose no time in starting their profit making scheme, and this time too they have surfaced in the area. Our correspondent from Kurigram writes that against the BADC's fixed rate of Tk. 180 for a 10-kg bag of boro seed, black marketers are selling it for about Tk. 230. While this has been going on quite in the open in the district, BADC authorities, as usual, have been denying any such corruption. We are surprised to further learn that some officials have even denied any seed crisis at the sales centres claiming that sufficient quantity of seeds would reach there within hours.:

Seed Crisis hits Boro (HYV Rice) Farming

Crisis of seeds is hampering cultivation of Boro paddy in the district in the current season. Farmers of the district are facing problems in procuring Boro seeds from Bangaladesh

Agriculture Development Corporation (BADC). Despite remaining standing in the queue for long while the farmers are returning home without seeds from BADC sales centres. This correspondent visited BADC's seed sales centre in Kurigram recently and saw farmers standing in a long queue to procure seeds.

A 10-kg bag of Boro seed is being sold between Tk 205 and Tk 230 in the market against the government rate of Tk 180, they alleged. Besides, some dealers are selling seeds in the black market at exorbitant price as BADC authorities have failed to provide the farmers with seeds in time. As a result, the genuine farmers are being deprived of seeds, they added. Hafizur Rahman, sub-assistant director (seed) of Kurigram BADC's sales centre, denied crisis of seeds at the centre and said that seeds would reach the godown within a few hours.

On the other hand, Mohiuddin Ahmed, Deputy Director of BADC, said that a total of 250 tonnes of seeds are required for cultivating 70,000 hectares of land in the nine upazilas of the district this season. But BADC does not have sufficient quantity of seeds at its godown. The BADC authorities have appointed 34 dealers for selling seeds to the growers in the nine upazilas of the district, but only three dealers have opened their sales centres, he added (The Independent November 25, 2003).

Narrowly focusing on increasing production-as the Green Revolution does-cannot alleviate hunger because it fails to alter the tightly concentrated distribution of economic power, especially access to land and purchasing power. Even the World Bank concluded in a major 1986 study of world hunger that a rapid increase in food production does not necessarily result in food security-that is, less hunger. Current hunger can only be alleviated by "redistributing purchasing power and resources toward those who are undernourished," the study said. In a nutshell-if the poor don't have the money to buy food, increased production is not going to help them.

Is it true that the best way to fight hunger, protect the environment and reduce poverty in Africa is by relying on Green Revolution crop varieties, and using more imported farm chemicals, plus genetic engineering and free trade? This is precisely what powerful institutions in Washington and elsewhere are prescribing for the continent, yet each of these elements could actually worsen, rather than improve conditions for Africa's poor majority (Institute of Food and Development Policy and the University of California at Berkeley, USA, 2002).

Agricultural Development Corporation (BADC) introduced modern irrigation equipment, chemical fertilisers, pesticides, and new varieties of seed that was developed by the international rice research institutes to increase food production in Bangladesh. Thus seeds developed in international research stations were made available to farmers for dry season (boro) crops in 1968 and wet season (aman) crops in 1970. Farmers experience and advise were never consulted, while introducing modern irrigation system and chemical based agriculture, as they are regarded as"ignorant farmers". Introduction and implementation of "Seed-Fertiliser-Water" technology was financed by the World Bank, Asian Development Bank, IMF, industrial countries as credit and donation.

In Bangladesh traditional seeds are not available, and the poor farmers under bureaucratic and corrupt system are living near famine situation.

Acute Boro Seed Shortage in Nilphamari

Nov 28: An acute shortage of Boro seed is prevailing in the district, creating grave concern among the farmers. According to a source in the Agriculture Extension Department (AED) only 90 metric tons of Boro seeds were sanctioned by the BADC against the total demand of 27,328 metric tons in the district so far. The source revealed that this year the farmers of the district began to make their seedbeds in order to cultivate Boro paddy on 68.320 hectares of land. A total of 27,328 metric tons of Boro seeds are needed for it.

Every day the farmers are seen moving to and fro in collecting seeds.At present, the seeds of 10 kgs packet produced by the BADC are being sold at Tk 250 to Tk 300 in different hats and bazars in the district, whereas the rate of the Boro seed of 10 kgs packet produced by the BADC was fixed at Tk 170 for sale. When contacted, the Deputy Director of the district Agriculture Extension Department regarding the severe Boro seed crisis prevailing in the entire district, he admitted the reality(The Independent, November 29, 2004):

1. Serious Seeds Crisis
2. VALUE OF DIVERSITY
3. Botanical Garden needs care and protection.

Agriculture began between ten and fifteen thousand years ago through the efforts of hundreds of thousands of people on several continents, and in many different social and ecological situations. The agriculture they established, suited to their own needs, was developed over the course of several thousand years."

When human beings initially began to domesticate plants, they selected plants that had traits which ultimately would be of great advantage to people. For example, they collected seeds which were larger, matured at harvest time and were "non-shattering," meaning the seeds were not easily dispersed but clung to plants.

The repeated selection and sowing of these seeds led to the development of plants more and more amenable to cultivation. But the multiplicity of places in which this process occurred ensured that a wide variety of plants, adapted to the particular requirements and situations of people all over the world, would flourish. Today, however, the international genetics-supply industry--made up of multinational chemical, pharmaceutical and food companies--is absorbing the world's small and medium-sized seed companies and strengthening its control over the world's crops.

In promoting their narrow line of seed varieties, the multinationals are rapidly pushing the industrialized countries toward an overall sameness in their food supply and threatening the world's genetic diversity.

At the same time, they are expanding their sales to farmers in the Third World, the origin of most of the world's indispensable plant varieties. By allowing this to occur, the world is inviting environmental disaster and widespread hunger.

Farmers who begin to rely on the high-yield seeds sold by the highly concentrated international genetics supply industry soon find themselves caught in a trap. Uniform crops are especially vulnerable to pests and disease, forcing growers to use pesticides, often sold by the same companies that provide seeds. Consequently, individual farmers become increasingly dependent on the genetics-supply industry. The integration of independent farmers into the worldwide corporate agribusiness system also has a negative effect on biodiversity. In expressing general alarm over the fate of the more traditional seed varieties

Fowler and Mooney stress that the extinction of a seed variety does not come simply at the time when there are no more seeds; rather extinction comes when the seeds' development process ceases to exist. As farmers become more dependent on the genetics supply industry, they lose the ability to nurture that development process.

Fowler and Mooney point out that approximately 97 percent of the varieties on a 1903 USDA vegetable list are now extinct. The consequences of such uniformity were demonstrated by the 1970 corn blight, which destroyed over 15 percent of the U.S. crop. In the wake of the corn blight, a 1972 U.S. National Academy of Sciences report, Genetic Vulnerability of Major Crops, revealed that the United States was shockingly reliant on a handful of seed varieties for its major crops. Fowler and Mooney emphasize that the study concluded that U.S. agriculture was "impressively uniform genetically and impressively vulnerable." Tracing the origins of Latin America's coffee industry illustrates how dependent whole nations and regions of the world have become on a few crop varieties

The entire coffee industry of Latin America is based on the seven plants taken from Yemen a thousand years ago and then from one plant in Indonesia almost 300 years ago. Today, Fowler and Mooney report, there is little genetic diversity in coffee crops outside of Ethiopia.

But the Ethiopian government, which believes the country has historically not received fair compensation for its genetic resources, is refusing to permit future collection of coffee resources from Ethiopia. "Unless Ethiopia relents," Fowler and Mooney write, "we will have an opportunity to see what happens to a narrowly-based crop like coffee without recourse to badly needed genetic resources.

The Struggle among Corporate Agribusiness Interests

Third World governments, the United Nations Food and Agriculture Organization (FAO) and the International Board for Plant Genetic Resources (IBPGR) to establish ground rules that will ensure the preservation and exchange of germplasm among all countries.

Unfortunately, they report, international efforts at seed preservation have favoured crops of primary interest to breeders in the industrialized countries. As founding staff members of the Rural Advancement Foundation International (RAFI), a nonprofit organization working for just and sustainable agriculture, both Fowler and Mooney have been participants in, and eyewitnesses to, many of those struggles.

Based on their years of observations and careful research, the authors have sought to develop a constructive approach to the enormous political, economic and scientific problems that the questions of genetic diversity pose. They have concluded that following principles of genetic conservation are necessary to preserve a healthy and diversified seed culture:

- Decentralized, broad-based strategies of preservation.
- Multiple strategies must be employed, with as many groups as possible--including not just scientists, but farmers, fishing people and medicine makers--consulted about what should be saved and how.
- Agricultural diversity must be used, so that it continues to evolve and retain its value.
- And, efforts to save agricultural diversity must be interwoven with efforts to save the farm community.

"Diversity", like music or a dialect, is part of the community that produced it. It cannot exist for long without that community and the circumstances that gave rise to it. Saving farmers is a prerequisite of saving diversity.

Conversely, communities must save their agricultural diversity in order to retain their own options for development and self-reliance. Someone else's seeds imply someone else's needs." "We each have a special role to play in passing this gift [genetic diversity] on to the next generation," write Fowler and Mooney. Throughout history, they show, it has been "amateurs"--people who love their seeds--more than scientists who have saved diversity.. If biodiversity is cared well in a country, the nation could be benefited by producing optimum harvest of its bioresources. Biodiversity then sponsors the sustenance of economic development of nation. So, we care biodiversity for two main causes for our existence in the world: a. We care for scientific/environmental reasons; and b. We care for commercial/ trade reasons.

We care biodiversity for scientific/environmental reasons because it has great role in nature on the following headings: 1. Keeps the balance of biomass production in the biosphere; 2. Keeps the energy-flow in biotic form in balance in biosphere; 3. Keeps the soil fertility in balance in biosphere (eg. Keeps the proportionity of production of microbes in respective areas in the biosphere); and 4. Keeps the biotic-abiotic relations in balance (Ecosystem balance maintenance).

HEREDITY AND THE GENE

Knowledge of the principles of heredity is so basic to our fundamental understanding of the biological sciences that it is hard to believe that these principles were discovered only in the 1860s (and their importance was realized less than a century ago). However, a practical knowledge of the hereditary process came long before its mechanism was understood. Archeologists have discovered that as long as 7,000 years ago farmers in Central America were improving crops of corn by planting hybrid corn seeds that had developed preferred characteristics. Over 6,000 years ago, the Chinese learned how to develop superior strains of rice.

An ancient Babylonian tablet shows a pedigree of a family of horses through five generations, with detailed information about height, length of the mane, and other traits, revealing that they had some knowledge that these traits were transmitted. Farmers and gardeners have continued to practice this type of selective breeding in both plants and animals. Each time an individual plant or animal appeared with a desired characteristic, it was bred again to produce more with similar traits. For example, at harvest time farmers would select heads of wheat that had the most or largest kernels and save them to use as seed the next year.

Science (or more generally, curiosity) received a large boost during the Renaissance. During the 16th and 17th centuries, the interest in the physical and natural sciences grew. This increased interest laid the foundation for science as we know it now. As gene technology bears on a large number of different disciplines (biochemistry, genetics, cell biology, engineering, etc.), a couple of key findings in different areas will be discussed here. It is interesting to note that some of the great discoveries that were made were not viewed to be very important at the time, just because the proper framework to understand the importance of the findings had not yet been developed. Also, it is important to realize that many of the seminal discoveries were made at the interface between two or more disciplines. The importance of this interdisciplinary insight and research toward developing new concepts and creating new ways of looking at problems persists to this day.

By the end of the 17th century, Hooke in England and van Leeuwenhoek in The Netherlands came up with microscopes. The microscope constructed by van Leeuwenhoek magnified 200 times, enough to see into "a new world" of little beasties that no one had ever seen. Hooke observed how cork was composed of a series of tiny cubicles, much like a honeycomb. He coined the word "cell" for such a cubicle, since it reminded him of the tiny cubicles used by monks in monasteries.

In the beginning of the 19th century, Brown, a Scottish botanist, found "opaque spots" in the cells of orchids. Upon further investigation, he found such spots in cells of all plants. He named the spot in the cell the nucleus, the latin name for "little nut". We will see later that the nucleus is the place where the genetic material is located and its expression is regulated. The current connotation of the word nucleus accurately represents the later realization that the nucleus is the "main office" of the cell, and has replaced the original meaning of the word.

It is striking that the discovery of the principles of heredity was not made by an eminent scientist, but rather by a monk doing experiments in the vegetable garden of a monastery. The monk, Gregor Mendel, was intrigued by the multitude of shapes and colors of all living things, even occurring within a single species growing/living at a single site. He used 34 different varieties of self-fertilizing peas for his experiments (self-fertilizing means that the egg under natural conditions is fertilized by the pollen from the same plant; in peas the flower remains closed during the fertilization period, thus preventing pollen from other plants fertilizing the eggs); he grew these varieties of peas for several pea generations.

He crossbred different kinds of peas by opening up a flower, cutting off the stamen, and putting pollen from another flower on the stigma. He tied a tiny bag over the flower to prevent any fertilization by other pollen. He worked on the problem for about eight years, and got results from 10,000 different plants. He selected seven characteristics to study (the color and the shape of the seed, the color of the flowers, the color and form of the pods, the position of the flowers, and the length of the stem), and systematically focused on one or two parameters at a time. In his first experiment, he crossbred plants with round and wrinkled seeds.

The progeny only had round seeds, which was in contradiction with the theory current at that time, that the progeny contained a blend of the characteristics of the two parents! He grew plants from the progeny seed, wondering what the seeds of these plants would look like. This time, there were both wrinkled and round seeds, the round seeds outnumbering the wrinkled ones by a factor of three. In the next generation, the plants grown from wrinkled seeds only gave wrinkled seeds, whereas plants from round seeds once again produced a mix of round and wrinkled ones, in a ratio of three to one. He did similar experiments while looking at the other properties he had selected.

On the basis of the experiments he did, he formulated several rules: (1) heredity is determined by distinct elements contained in the two cells contributed by the parents of the organism, and these elements combine randomly; (2) each characteristic of an organism is determined separately from the others; and (3) when two different characteristics combine, one characteristic will dominate over the other (for his second and third rule he just had luck; it turned out to be not applicable in all cases). He decided to use single letters for each characteristic; a capital letter denoted the dominant factor (the one that determines the phenotype), and the same letter in lower case denoted the recessive one (the one that does not express itself in the phenotype, unless the dominant factor is absent). He realized that progeny combined factors from both parents, and that in the pollen or the egg only half of the information of the parent was present.

This insight laid the foundation for modern genetics. He presented his findings and conclusions to a conference of local intellectuals. The response was a polite silence. Nevertheless, they invited him to publish his results in a scientific journal, which he did in 1866. His paper was totally neglected by the scientists of his time. He died 18 years later as a well-respected abbot, but without recognition for his pioneering work.

Mendel's pioneering work chronologically was paralleled by breakthroughs in biochemistry and cell anatomy. In 1869, a Swiss graduate student, Friedrich Miescher, set out to do a chemical analysis of a cell. He was having trouble breaking down all parts of the cell when he got the idea of "digesting" it with stomach enzymes. Cells broke down upon digestion, but they always left a residue from the nucleus. Miescher called the residue "nuclein". He combined pleasure with business angling for salmon in the Rhine river near his home. He had found that salmon sperm had very large nuclei, almost half of the cell itself, and Miescher was able to accumulate a good supply of his newly discovered nuclein.

It was fortunate that he did his experiments more than a century ago, because one would be hard-pressed to find any salmon in the polluted Rhine river now. One of Miescher's coworkers observed that nuclein was composed of carbon, oxygen, hydrogen, nitrogen and phosphorus. Jars and jars of the white sticky stuff, the nuclein, stood on Miescher's laboratory shelves for years, but he did not realize that what he had isolated was a crude preparation of DNA, the genetic material that contained the molecular basis for Mendel's results. The chemical characterization of the nuclein was published by Miescher's friends after his death, but at that time it was just a curiosity, a piece of the puzzle that could not be fitted into its proper perspective.

In tbe middle of the 19th century, a young lad by the name of William Perkin was asked by his employer to find ways to synthetically make quinine, a natural drug used to treat malaria and other maladies. He went ahead, and mixed all kinds of chemicals. One mix, aniline and potassium dichromate resulted in a purple goo. Adding alcohol made a wonderful purple solution that did not cure malaria, but could be used as a dye. A quarter of a century later, a German scientist, Walter Flemming, stained some cells with Perkin's dye, to see how that would look under the microscope. Suddenly, the nucleus Brown had described half a century earlier became distinctly clear, since it absorbed the dye.

Flemming called the mass of colored material in the nucleus chromatin. When he stained sections of growing tissue, he saw the cells in the process of division. He watched the chromatin bunch up into short, threadlike bundles, and he called these chromosomes. As he studied the bunching up process, he saw the chromosomes double in number. He watched the chromosomes begin to pull apart. Half of them moved to one end of the cell, the other half went to the other end. Because the chromosomes spread out like threads during this process, Flemming called the process mitosis ("thread" in Greek). Half the chromosomes moved into one of the new daughter cells that formed when the cell divided, and half went into the other new cell. They were exact duplicates. When Flemming wrote about mitosis and chromosomes in 1882, he did not connect it to inheritance. Mendel's laws had not yet been rediscovered.

Five years later, van Beneden, a Belgian scientist, discovered that every cell in a body has the same number of chromosomes, with the exception of the sperm or egg cell, which contains only half the "usual" number of chromosomes. He also found that different species have different numbers of chromosomes: a human has 46, a fruit fly has 8, and a crayfish has some 200. The giant sequoia can live with 22. Note that up to this point all important findings in this area had been made in Europe. Basic sciences in the US had not come of age yet.

Around the beginning of the 20th century, three scientists (De Vries in The Netherlands, Correns in Germany, and Tschermak in Austria) independently and virtually simultaneously performed plant breeding experiments that greatly resembled the work done by Mendel half a century earlier. All three only discovered Mendel's publication while doing a literature study in preparation of their publication. Around the same time, the name "gene" was coined to describe the hereditary substance, and "genetics" became the name of the rapidly emerging discipline of gene study and characterization.

Advances in the field of genetics were also fueled by the discoveries by Flemming, van Beneden, and others on the cellular/sub-cellular level. In the twenties and thirties of this century, chromosomes became the subject of intense study, and it is mainly the pioneering work of two groups (those of Morgan and Castle) that has contributed greatly to the further development of the concept of genes in relation to chromosomes. Both groups studied the fruitfly (Drosophila melanogaster), whose eight chromosomes are all readily visible under the light microscope. Six chromosomes could be divided up in three pairs of morphologically identical chromosomes, whereas the other two (named X) were identical in females while in males one of the two was replaced by a much smaller chromosome (Y).

It was correctly assumed earlier (from work on grasshoppers by Walter Sutton) that these two chromosomes would be the ones that determined the sex of the organism. Morgan found a fruitfly mutant which had white eyes rather than the usual red ones. When this mutant, a male, was mated with a red-eyed female, all the progeny had red eyes, indicating that the red eyes were dominant. Upon self-breeding of the progeny, the red-eyed flies outnumbered the white-eyed ones by a factor of about three, as would be expected. However, it was found that all white-eyed flies were male. Thus, according to Morgan, the gene for eye color must be on the X chromosome. He was right. This was the first time that anyone had placed a specific gene on a specific chromosome.

Of course, Morgan could not do all this work by himself. He had a large group of graduate students (nicknamed the "fly squad"), some of whom later became independent researchers. One of them was Hermann Muller. He was wondering what caused the mutations. He tried outside changes: no luck: if you cut off the tail of a mouse, all progeny had tails. He tried accidents of all kinds, but no mutations were produced. He then began thinking in terms of the "world of the little". Perhaps mutations came from an ultramiscroscopic accident. Maybe the gene could not escape a speeding electron. He then tried radiation (small, energy-rich particles); he reasoned that there is some natural radiation in the atmosphere, causing rare mutations, and that he might be able to get many more mutations by irradiating his flies with X-rays (invented a couple of years before). He put hundreds of flies in gelatin capsules, irradiated them, and then bred them with untreated flies. The progeny was an interesting mix: there were flies with bulging eyes, flat eyes, dented eyes, purple eyes, yellow eyes, broad wings, curly wings, bumpy wings, etc.. Some were hyperactive, others dopy. Some were big, others small. The conclusion: mutations are caused by ultramiscroscopic collisions, making changes in the hereditary information.

The question now could focus on the chemical identity of the carrier of the hereditary information. As indicated earlier, Miescher had isolated nuclein in large amounts from salmon sperm, and unlike protein, nuclein was found to have a high concentration of phosphorus. Therefore, it was unlikely that nuclein was just a type of protein. Eventually, chemical analysis revealed that chromosomes are made of both protein and nuclein, and that nuclein is deoxyribonucleic acid (DNA). With both protein and DNA present in chromosomes, which of these carries hereditary information? Almost everyone thought that proteins must be the hereditary material because chemically it is much more complicated than DNA, and the molecule of heredity had to be able to contain an extraordinary amount of information. DNA seemed too simple..... How wrong this reasoning would turn out to be.

In 1928 Frederick Griffith, a scientist who was interested in pathology rather than in genes, made a serendipitous discovery that led to the identification of the chemical composition of genetic material. He was trying to understand the differences between the strain of Diplococcus pneumoniae that cause pneumonia (virulent strains) and the strains that are harmless (non-virulent). The virulent forms had polysaccharide capsules around them that gave them a smooth appearance. The non-virulent bacteria had no capsules and were rough. Griffith hoped that either heat-killed virulent strains or live non-virulent strains could be used as vaccine. He mixed heat-killed virulent and live non-virulent strains together in hopes of making an effective vaccine. Instead, upon injecting this mixture into mice, the mice died. How could dead bacteria be virulent? He retrieved virulent bacteria from the blood of the dead mice, and he saw that the bacteria were smooth. Something had been transferred from the dead virulent bacteria to the live non-virulent ones. Griffith called this substance the "transforming factor". However, it is unfortunate that Griffith is not given any credit at the MIT web site, and only the people who followed up on it (Avery et al.) are mentioned there.

Griffith did not follow up on identifying the chemical nature of his "transforming factor". This work laid dormant for a while until 1943, when O.T. Avery and colleagues purified the "transforming factor" and found that it was DNA. This finding was viewed with a lot of suspicion by other scientists, as everyone thought that protein (consisting of a long string of subunits, each of which could have 20 different chemical structures) must be the location of hereditary information. DNA (consisting of a much longer string of subunits, each of which could have 4 different chemical structures) was viewed as inadequate to store much information. This way of thinking is typically B.C. (before computers) because we know now that an enormous amount of information can be stored even using a binary (on/off or 0/1) mode.

Developments were rapid from then on. In 1952 the DNA-vs-protein debate was settled conclusively by showing using radioactive isotopes that protein is not passed on from a virus to its progeny, but DNA is. The polynucleotide (DNA) molecule has a number of relatively unusual characteristics for a macromolecule, including the repeated negative charge of the phosphate groups, and the innate affinity of pyrimidine and purine bases to each other. Erwin Chargaff discovered in the forties that the base compositions of DNA from different organisms vary over a relatively wide range (for example, adenine makes up between 23 and 32% of the bases), but that the amount of A consistently was very close to the amount of T, and the amount of C essentially identical to that of G. These two equalities were the first indication that stoichiometric complexation occurs between A and T and between G and C.

Chargaff shared his findings with James Watson, who with the help of molecular models discovered that H-bonded base-paired structures could be formed between A and T, and between G and C, and that these structures have the same overall dimensions. Two H-bonds can be formed between A and T, and three between G and C. Watson brought this information to the attention of Francis Crick, a crystallographer. Methods had been developed to determine the structure of crystals (repetitively arranged groups of molecules) to atomic resolution, and Crick was an expert in using this technique. This culminated in the elucidation of the DNA structure in 1953.

The X-ray diffraction pattern of crystallized DNA could be interpreted in terms of a helix (essentially a winding staircase) composed of two polynucleotide strands with H-bonded pairs formed between the bases of opposing strands. Molecular-modeling studies showed that the H-bonded base pairs could form only when the directional senses of the two interacting chains were opposite (anti-parallel). Also, the DNA duplex was found to be "folded" in a three-dimensional structure, a double-helix (like two helices folded into each other). Watson and Crick Crick (but not Chargaff... life not always is fair) got the Nobel Prize in 1962. The Nobel Committee noted that the discovery had "no immediate practical application, but determining the molecular structure of the substance that is responsible for the forms that life takes is a discovery of tremendous importance". The understatement of the year? Without some of these breakthroughs, molecular biology and biotechnology might still be in their infancy.

ANTIBIOTICS AND PROTEINS MADE BY BACTERIA

The rapidly increasing resistance of selected microbes against a large number of antibiotics has become a very serious issue worldwide.

Antimicrobial Resistance

Since their discovery during the 20th century, antimicrobial agents (antibiotics and related medicinal drugs) have substantially reduced the threat posed by infectious diseases. The use of these "wonder drugs", combined with improvements in sanitation, housing, and nutrition, and the advent of widespread immunization programmes, has led to a dramatic drop in deaths from diseases that were previously widespread, untreatable, and frequently fatal. Over the years, antimicrobials have saved the lives and eased the suffering of millions of people. By helping to bring many serious infectious diseases under control, these drugs have also contributed to the major gains in life expectancy experienced during the latter part of the last century. These gains are now seriously jeopardized by another recent development: the emergence and spread of microbes that are resistant to cheap and effective first-choice, or "first-line" drugs. The bacterial infections which contribute most to human disease are also those in which emerging and microbial resistance is most evident: diarrhoeal diseases, respiratory tract infections, meningitis, sexually transmitted infections, and hospital-acquired infections. Some important examples include penicillin-resistant Streptococcus pneumoniae, vancomycin-resistant enterococci, methicillin-resistant Staphylococcus aureus, multi-resistant salmonellae, and multi-resistant Mycobacterium tuberculosis. The development of resistance to drugs commonly used to treat malaria is of particular concern, as is the emerging resistance to anti-HIV drugs.

CONSEQUENCES

The consequences are severe. Infections caused by resistant microbes fail to respond to treatment, resulting in prolonged illness and greater risk of death. Treatment failures also lead to longer periods of infectivity, which increase the numbers of infected people moving in the community and thus expose the general population to the risk of contracting a resistant strain of infection.

When infections become resistant to first-line antimicrobials, treatment has to be switched to second-or third-line drugs, which are nearly always much more expensive and sometimes more toxic as well, e.g. the drugs needed to treat multidrug-resistant forms of tuberculosis are over 100 times more expensive than the first-line drugs used to treat non-resistant forms. In many countries, the high cost of such replacement drugs is prohibitive, with the result that some diseases can no longer be treated in areas where resistance to first-line drugs is widespread. Most alarming of all are diseases where resistance is developing for virtually all currently available drugs, thus raising the spectre of a post-antibiotic era. Even if the pharmaceutical industry were to step up efforts to develop new replacement drugs immediately, current trends suggest that some diseases will have no effective therapies within the next ten years.

CAUSES

Microbes (the collective term for bacteria, fungi, parasites, and viruses) cause infectious diseases, and antimicrobial agents, such as penicillin, streptomycin, and more than 150 others, have been developed to combat the spread and severity of many of these diseases. Resistance to antimicrobials is a natural biological phenomenon that can be amplified or accelerated by a variety of factors, including human practices. The use of an antimicrobial for any infection, real or feared, in any dose and over any time period, forces microbes to either adapt or die in a phenomenon known as "selective pressure". The microbes which adapt and survive carry genes for resistance, which can be passed on.

Bacteria are particularly efficient at enhancing the effects of resistance, not only because of their ability to multiply very rapidly but also because they can transfer their resistance genes, which are passed on when the bacteria replicate. In the medical setting, such resistant microbes will not be killed by an antimicrobial agent during a standard course of treatment. Resistant bacteria can also pass on their resistance genes to other related bacteria through "conjugation", whereby plasmids carrying the genes jump from one organism to another. Resistance to a single drug can thus spread rapidly through a bacterial population. When anti-microbials are used incorrectly-for too short a time, at too low a dose, at inadequate potency; or for the wrong disease-the likelihood that bacteria and other microbes will adapt and replicate rather than be killed is greatly enhanced. Much evidence supports the view that the total consumption of antimicrobials is the critical factor in selecting resistance. Paradoxically, underuse through lack of access, inadequate dosing, poor adherence, and substandard anti-microbials may play as important a role as overuse. For these reasons, improving use is a priority if the emergence and spread of resistance are to be controlled.

UNPRECEDENTED TRENDS

In the past, medicine and science were able to stay ahead of this natural phenomenon through the discovery of potent new classes of antimicrobials, a process that flourished from 1930-1970 and has since slowed to a virtual standstill, partly because of misplaced confidence that infectious diseases had been conquered, at least in the industrialized world. In just the past few decades, the development of resistant microbes has been greatly accelerated by several concurrent trends. These have worked to increase the number of infections and thus expand both the need for antimicrobials and the opportunities for their misuse. Such trends include:

- urbanization with its associated overcrowding and poor sanitation, which greatly facilitate the spread of such diseases as typhoid, tuberculosis, respiratory infections, and pneumonia;
- pollution, environmental degradation, and changing weather patterns, which can affect the incidence and distribution of infectious diseases, especially those, such as malaria, that are spread by insects and other vectors;
- demographic changes, which have resulted in a growing proportion of elderly people needing hospital-based interventions and thus at risk of exposure to highly resistant pathogens found in hospital settings;
- the AIDS epidemic, which has greatly enlarged the population of immunocompromised patients at risk of numerous infections, many of which were previously rare;
- the resurgence of old foes, such as malaria and tuberculosis, which are now responsible for many millions of infections each year;
- the enormous growth of global trade and travel which have increased the speed and facility with which both infectious diseases and resistant microorganisms can spread between continents.

As the number of infections and the corresponding use of antimicrobials have increased, so has the prevalence of resistance. In addition, the enhanced food requirements of an expanding world population have led to the widespread routine use of antimicrobials as growth promoters or preventive agents in food-producing animals and poultry flocks. Such practices have likewise contributed to the rise in resistant microbes, which can be transmitted from animals to man.

FACTORS THAT ENCOURAGE THE SPREAD OF RESISTANCE

The emergence and spread of antimicrobial resistance are complex problems driven by numerous interconnected factors, many of which are linked to the misuse of antimicrobials and thus amenable to change. In turn, antimicrobial use is influenced by an interplay of the knowledge, expectations, and interactions of prescribers and patients, economic incentives, characteristics of a country's health system, and the regulatory environment.

Patient-related factors are major drivers of inappropriate antimicrobial use. For example, many patients believe that new and expensive medications are more efficacious than older agents. In addition to causing unnecessary health care expenditure, this perception encourages the selection of resistance to these newer agents as well as to older agents in their class. Self-medication with antimicrobials is another major factor contributing to resistance. Self-medicated antimicrobials may be unnecessary, are often inadequately dosed, or may not contain adequate amounts of active drug, especially if they are counterfeit drugs. In many developing countries, antimicrobials are purchased in single doses and taken only until the patient feels better, which may occur before the pathogen has been eliminated. Inappropriate demand can also be stimulated by marketing practices. Direct-to-consumer advertising allows pharmaceutical manufacturers to market medicines directly to the public via television, radio, print media, and the Internet. In particular, advertising on the Internet is gaining market penetration, yet it is difficult to control with legislation due to poor enforceability.

Prescribers' perceptions regarding patient expectations and demands substantially influence prescribing practice. Physicians can be pressured by patient expectations to prescribe antimicrobials even in the absence of appropriate indications. In some cultural settings, antimicrobials given by injection are considered more efficacious than oral formulations. Such perceptions tend to be associated with the over-prescribing of broad-spectrum injectable agents when a narrow-spectrum oral agent would be more appropriate. Prescribing "just to be on the safe side" increases when there is diagnostic uncertainty, lack of prescriber knowledge regarding optimal diagnostic approaches, lack of opportunity for patient follow-up, or fear of possible litigation. In many countries, antimicrobials can be easily obtained in pharmacies and markets without a prescription. Patient compliance with recommended treatment is another major problem. Patients forget to take medication, interrupt their treatment when they begin to feel better, or may be unable to afford a full course, thereby creating an ideal environment for microbes to adapt rather than be killed. In some countries, low quality antibiotics (poorly formulated or manufactured, counterfeited or expired) are still sold and used for self-medication or prophylaxis.

Hospitals are a critical component of the antimicrobial resistance problem worldwide. The combination of highly susceptible patients, intensive and prolonged antimicrobial use, and cross-infection has resulted in nosocomial infections with highly resistant bacterial pathogens. Resistant hospital-acquired infections are expensive to control and extremely difficult to eradicate. Failure to implement simple infection control practices, such as handwashing and changing gloves before and after contact with patients, is a common cause of infection spread in hospitals throughout the world. Hospitals are also the eventual site of treatment for many patients with severe infections due to resistant pathogens acquired in the community. In the wake of the AIDS epidemic, the prevalence of such infections can be expected to increase.

Veterinary prescription of antimicrobials also contributes to the problem of resistance. In North America and Europe, an estimated 50% in tonnage of all antimicrobial production is used in food-producing animals and poultry. The largest quantities are used as regular supplements for prophylaxis or growth promotion, thus exposing a large number of animals, irrespective of their health status, to frequently subtherapeutic concentrations of antimicrobials. Such widespread use of antimicrobials for disease control and growth promotion in animals has been paralleled by an increase in resistance in those bacteria (such as Salmonella and Campylobacter) that can spread from animals, often through food, to cause infections in humans.

THE NEED FOR A GLOBAL RESPONSE

In September 2001, WHO launched the first global strategy for combating the serious problems caused by the emergence and spread of antimicrobial resistance. Known as the WHO Global Strategy for Containment of Antimicrobial Resistance, the strategy recognizes that antimicrobial resistance is a global problem that must be addressed in all countries. No single nation, however effective it is at containing resistance within its borders, can protect itself from the importation of resistant pathogens through travel and trade. Poor prescribing practices in any country now threaten to undermine the potency of vital antimicrobials everywhere.

The strategy recommends interventions that can be used to slow the emergence and reduce the spread of resistance in a diverse range of settings. The interventions are organized according to groups of people whose practices and behaviours contribute to resistance and where changes are judged likely to have a significant impact at both national and international levels. These include consumers, prescribers and dispensers, veterinarians, and managers of hospitals and diagnostic laboratories as well as national governments, the pharmaceutical industry, professional societies, and international agencies. Global principles for the containment of antimicrobial resistance in food-producing animals were issued by WHO in June 2000. As much of the responsibility for containing resistance rests with national governments, the strategy gives particular attention to interventions involving the introduction of legislation and policies governing the development, licensing, distribution, and sale of antimicrobial agents. The strategy is sufficiently flexible to be applied in poor and wealthy nations alike. The process for selecting the necessary interventions to limit emerging antimicrobial resistance can be based on the diseases most prevalent in a given country. In advocating widespread adoption of this strategy, WHO aims to encourage the urgent actions needed to reverse or at least curtail trends which have major economic as well as health implications. Moreover, in view of the global nature of the antimicrobial resistance problem, the efforts of any nation to implement the WHO Global Strategy are likely to be felt worldwide.

The strategy builds on a number of WHO activities aimed at both monitoring the global emergence and spread of antimicrobial resistance and extending direct support to countries. WHO helps countries establish laboratory-based networks for the surveillance of resistance. Specific activities include staff training, support in methods for the quality assurance of laboratory tests, and provision of laboratory reagents. In addition, WHO distributes a computer software program, WHONET. Microbiologists, clinicians, and infection control workers may use this software to improve the systematic monitoring of drug resistance in their hospitals and communities and to share their data in a common format among national networks.

Since 1977, WHO has produced Model Lists of Essential Drugs in order to help governments select the most effective and appropriate drugs in line with priority needs. The lists, which are regularly revised, also contribute to the rational purchasing and use of drugs. Studies have demonstrated that in those areas in which an essential drugs programme is in operation, significantly more essential drugs are available, significantly fewer injections and antimicrobials are utilized, and drug stocks last about three times longer than in regions without such a programme. At present over 120 countries have implemented an essential drugs list. With the first global strategy for containment of antimicrobial resistance now available, WHO is also in a position to advise health policy-makers and managers on the specific interventions needed to safeguard the effectiveness of vital drugs and thus ensure that their life-saving capacity remains available to future generations.

QUESTIONS ABOUT ANTIBIOTIC RESISTANCE

What are bacteria and viruses?

Bacteria are single-celled organisms usually found all over the inside and outside of our bodies, except in the blood and spinal fluid. Many bacteria are not harmful. In fact, some are actually beneficial. However, disease-causing bacteria trigger illnesses, such as strep throat and some ear infections. Viruses are even smaller than bacteria. A virus cannot survive outside the body's cells. It causes illnesses by invading healthy cells and reproducing.

What kinds of infections are caused by viruses and should not be treated with antibiotics?

1. Colds
2. Flu
3. Most coughs and bronchitis
4. Sore throats (except for those resulting from strep throat)

How do I know when an illness is caused by a viral or bacterial infection?

Sometimes it is very hard to tell. Consult with your doctor to be sure.

When do I need to take antibiotics?

Antibiotics are very powerful medications. They should only be used when prescribed by a doctor to treat bacterial infections. Do I need an antibiotic when mucus from the nose changes to yellow or green?

Yellow or green mucus does not indicate a bacterial infection. It is normal for the mucus to get thick and change color during a viral cold.

Should I ask my doctor to prescribe antibiotics?

Talk to your doctor about the best treatment. You should not expect to get a prescription for antibiotics. If you have a viral infection, antibiotics will not cure it, help you feel better, or prevent someone else from getting your virus.

What is antibiotic resistance and why should I be concerned?

Antibiotic resistance occurs when bacteria change in a way that reduces or eliminates the effectiveness of antibiotics. These resistant bacteria survive and multiply-causing more harm, such as a longer illness, more doctor visits, and a need for more expensive and toxic antibiotics. Resistant bacteria may even cause death.

What can I do to avoid antibiotic-resistant infections?

Start by talking with your healthcare provider about antibiotic resistance.

Ask whether an antibiotic is likely to be effective in treating your illness.

Do not demand an antibiotic when your healthcare provider determines one is not appropriate.

Ask what else you can do to help relieve your symptoms.

What can I do to protect my child from antibiotic-resistant bacteria?

Use antibiotics only when your doctor has determined that they are likely to be effective. Antibiotics will not cure most colds, coughs, sore throats, or runny noses. Children fight off colds on tneir own.

If mucus from the nose changes from clear to yellow or green, does this mean that my child needs an antibiotic?

Yellow or green mucus does not mean that your child has a bacterial infection. It is normal for the mucus to get thick and change color during a viral cold.

Does this mean that I should never give my child antibiotics?

Antibiotics are very powerful medicines and should only be used to treat bacterial infections. If an antibiotic is prescribed, make sure you take the entire course and never save the medication for later use.

How do I know if my child has a viral or bacterial infection?

Ask your doctor. If you think that your child might need treatment, you should contact your doctor. But remember, colds are caused by viruses and should not be treated with antibiotics.

FACTS ABOUT ANTIBIOTIC RESISTANCE

Antibiotic resistance has been called one of the world's most pressing public health problems.

The number of bacteria resistant to antibiotics has increased in the last decade. Nearly all significant bacterial infections in the world are becoming resistant to the most commonly prescribed antibiotic treatments.

Every time a person takes antibiotics, sensitive bacteria are killed, but resistant germs may be left to grow and multiply. Repeated and improper uses of antibiotics are primary causes of the increase in drug-resistant bacteria. Misuse of antibiotics jeopardizes the usefulness of essential drugs. Decreasing inappropriate antibiotic use is the best way to control resistance.

Children are of particular concern because they have the highest rates of antibiotic use. They also have the highest rate of infections caused by antibiotic-resistant pathogens.

Parent pressure makes a difference. For pediatric care, a recent study showed that doctors prescribe antibiotics 65% of the time if they perceive parents expect them; and 12% of the time if they feel parents do not expect them.

Antibiotic resistance can cause significant danger and suffering for people who have common infections that once were easily treatable with antibiotics. When antibiotics fail to work, the consequences are longer-lasting illnesses; more doctor visits or extended hospital stays; and the need for more expensive and toxic medications. Some resistant infections can cause death.

PREVENTION OF ANTIBIOTIC RESISTANCE

Do not take an antibiotic for a viral infection like a cold, a cough or the flu.

Take an antibiotic exactly as the doctor tells you. Do not skip doses. Complete the prescribed course of treatment, even if you are feeling better.

Do not save any antibiotics for the next time you get sick. Discard any leftover medication once you have completed your prescribed course of treatment.

Do not take antibiotics prescribed for someone else. The antibiotic may not be appropriate for your illness. Taking the wrong medicine may delay correct treatment and allow bacteria to multiply.

Antibiotic prescriptions in outpatient settings can be reduced dramatically-without adversely affecting patient health-by not prescribing antibiotics for viral illnesses, such as colds, most sore throats, coughs, bronchitis, and the flu.

Parents should not demand antibiotics when a healthcare provider has determined they are not needed.

Parents should talk with their healthcare provider about antibiotic resistance. Parents should not give their children antibiotics for a viral infection like a cold, a cough, or the flu. Antibiotics should be used only to treat bacterial infections.

Parents should ensure that their children take all medication as prescribed, even if symptoms disappear. If treatment stops too soon, some bacteria may survive and re-infect.

A Prescription for Parents: Five Hints to Understanding Antibiotic Usage. When are antibiotics necessary? Your doctor can best answer this complicated question and the answer depends on the diagnosis. Here are a few examples:

1. *Ear Infections:* There are several types; many need antibiotics, but some do not.
2. *Sinus Infections:* Most children with thick or green mucus do not have sinus infections. Antibiotics are needed for some long-lasting or severe cases.
3. *Cough or Bronchitis:* Children rarely need antibiotics for bronchitis.
4. *Sore Throat:* Viruses cause most cases. Only one major kind, "strep throat," requires antibiotics. This condition must be diagnosed by a laboratory test.
5. *Colds:* Colds are caused by viruses and may last for two weeks or longer. Antibiotics have no effect on colds, but your doctor may have suggestions for obtaining comfort while the illness runs its course.

It is worth noting that viral infections sometimes lead to bacterial infections. But treating viral infections with antibiotics will not prevent bacterial infections and may trigger infections with resistant bacteria. Keep your doctor informed if the illness gets worse, or lasts a long time, so that the proper treatment can be given as needed.

Fluid in the Middle Ear

A doctor said your child has fluid in the middle ear, also called otitis (oh-TIE-Tus) media with effusion (uh-FEW-zhun)

(OME). Fluid usually does not bother children and it almost always goes away on its own. This does not have to be treated with antibiotics, unless it lasts for a few months. Here are some facts about OME and ear infections.

What are the main kinds of ear infections?

Swimmer's ear (otitis externa) is an infection of the ear canal that can be painful and is treated with eardrops.

A middle ear infection, which a doctor might call "acute otitis media" (AOM), may cause ear pain, fever, or an inflamed eardrum, and is often treated with oral antibiotics.

What causes OME?

Fluid may build up in the middle ear for two reasons. When a child has a cold, the middle ear makes fluid just as the nose does-it just doesn't run out as easily from the middle ear. After a middle ear infection, fluid may take a month or longer to go away.

Are antibiotics ever needed for OME?

Sometimes antibiotics may be needed if the fluid is still present after a few months and is causing decreased hearing in both ears. For this reason, your child will need an ear check in a few months. If there is still fluid in the middle ear, your child may need a hearing test.

What should I do?

The best treatment is to wait and watch your child. Since fluid in the middle ear rarely bothers children, it is best to let it go away on its own. Right now, your child might not need antibiotics.

You may need to schedule a visit to see the doctor again in a few months to be sure the fluid is gone.

Why not try antibiotics now?

Taking antibiotics when they are not needed can be harmful. Each time people take antibiotics, they are more likely to carry resistant germs in their noses and throats. Common antibiotics cannot kill these resistant germs. Your child may need antibiotics that are more costly, given by a needle, and/or administered in the hospital. Since OME will almost always get better on its own, it is better to wait and take antibiotics only when they are needed.

A Child's Runny Nose

Your child has a runny nose. This is a normal part of what happens during the common cold and as it gets better. Here are some facts about colds and runny noses.

What causes a runny nose during a cold?

When germs that cause colds first infect the nose and sinuses, the nose makes clear mucus. This helps wash the germs from the nose and sinuses.

After two or three days, the body's immune cells fight back, changing the mucus to a white or yellow color. As the bacteria that live in the nose grow back, they may also be found in the mucus, which changes the mucus to a greenish color. This is normal and does not mean your child needs antibiotics.

What should I do?

The best treatment is to wait and watch your child. Runny nose, cough, and symptoms like fever, headache, and muscle aches may be bothersome, but antibiotics will not make them go away any faster. Some people find that using a cool mist vaporizer or saltwater nose drops makes their child feel better.

Are antibiotics ever needed for a runny nose?

Antibiotics are needed only if your doctor tells you that your child has sinusitis. Your child's doctor may prescribe other medicine or give you tips to help with a cold's other symptoms like fever and cough, but antibiotics are not needed to treat the runny nose.

Why not try antibiotics now?

Taking antibiotics when they are not needed can be harmful. Each time people take antibiotics, they are more likely to carry resistant germs. Your child may need antibiotics that are more costly, given by a needle, and/or administered in the hospital. Since a runny nose almost always gets better on its own, it is better to wait and take antibiotics only when they are needed.

Cold and Flu Season

Colds, flu, and most sore throats and bronchitis are caused by viruses. Antibiotics do not help fight viruses. And they may do more harm than good: taking antibiotics when they are not needed-and cannot treat the illness-increases the risk of a resistant infection later.

Antibiotics Are Not for Colds and Flu.

Most infections are caused by two main types of germs-bacteria and viruses.

Bacteria are organisms found almost anywhere, except normally sterile sites, such as the blood stream and spinal fluid. A few bacteria, known as pathogens, can cause diseases in humans, animals, and plants.

Viruses are organisms that cause disease by invading healthy host cells. As virus particles multiply, the host cells burst, allowing the viruses to infect other cells.

COUGH AND COLD MEDICINES FOR CHILDREN

What can parents do if their children are too young or the healthcare provider advises against using cough and cold medicines?

Parents might consider clearing nasal congestion in infants with a rubber suction bulb. Also, secretions can be softened with saline nose drops or a cool-mist humidifier.

Are cough and cold medicines safe for children under 2 years of age?

There are no Food and Drug Administration (FDA)-approved dosing recommendations for children under 2 years of age. These drugs can, in rare cases, be harmful or even fatal. Parents and healthcare providers should use caution when giving cough and cold medicines to children under 2 years of age.

Do cough and cold medicines work in children under 2 years of age?

There is little evidence that cough and cold medicines work in children under 2 years of age.

Should parents give cough and cold medicines to children under 2 years of age?

Parents should consult a healthcare provider before giving cough and cold medicines to their children and should always tell providers about all prescription and over-the-counter medicines they are giving their child.

Should healthcare providers prescribe cough and cold medicines to children under 2 years of age?

Healthcare providers should exercise caution when recommending or prescribing cough and cold medicines to children under 2 years of age and should always ask caregivers about any other cough and cold medicines the child might be receiving. No FDA-approved dosing recommendations exist for over-the-counter cough and cold medicines in children under 2 years of age.

What should parents and doctors be careful of if they want to give cough and cold medicines to children under 2 years of age?

Be especially careful if giving more than one cough and cold medicine at a time to children under 2 years of age. Two medicines may have different brand names but may contain the same ingredient. Some cough and cold medicines contain more than one active ingredient.

ANTIBACTERIAL CLEANING AGENTS, ACNE MEDICATION, AND PROBIOTICSS

Q. Are antibacterial-containing products (soaps, household cleaners, etc.) better for preventing the spread of infection? Does their use add to the problem of resistance?

A. An essential part of preventing the spread of infection in the community and at home is proper hygiene. This includes hand-washing and cleaning shared items and surfaces. Antibacterial-containing products have not been proven to prevent the spread of infection better than products that do not contain antibacterial chemicals. Although a link between antibacterial chemicals used in personal cleaning products and bacterial resistance has been shown in in vitro studies, no human health consequence has been demonstrated. More studies examining resistance issues related to these products are needed.

The Food and Drug Administration (FDA) Nonprescription Drugs Advisory Committee voted unanimously on October 20, 2005 that there was a lack of evidence supporting the benefit of consumer products including handwashes, bodywashes, etc. containing antibacterial additives over similar products not containing antibacterial additives.

Q. Can antibiotic resistance develop from acne medication?

A. Antibiotic use, appropriate or otherwise, contributes to the development of antibiotic resistance. This is true for acne medications that contain antibiotics. Short and long-term use of antibiotics for treatment or prevention of bacterial infections should be under the direction of a physician to ensure appropriate use and detection of resistance.

Q. Do probiotics have a role in preventing or treating drug resistance or drug-resistant infections?

A. Probiotics are defined as microorganisms that when administered in sufficient quantities may improve health. There are a variety of probiotics that have been studied for various health benefits. Their role in preventing drug resistant infections in humans has not been established. CDC is currently monitoring research on probiotic use, but cannot make any recommendations at this time.

MRSA IN HEALTHCARE SETTINGS

MRSA has been featured in the news and on television programs a great deal recently. MRSA stands for Methicillin-resistant Staphylococcus aureus. This type of bacteria causes "staph" infections that are resistant to treatment with usual antibiotics. MRSA occurs most frequently among patients who undergo invasive medical procedures or who have weakened immune systems and are being treated in hospitals and healthcare facilities such as nursing homes and dialysis centers. MRSA in healthcare settings commonly causes serious and potentially life threatening infections, such as bloodstream infections, surgical site infections, or pneumonia. In addition to healthcare associated infections, MRSA can also infect people in the community at large, generally as skin infections that look like pimples or boils and can be swollen, painful and have draining pus. These skin infections often occur in otherwise healthy people.

How MRSA Spreads in Healthcare Settings

When we talk about the spread of an infection, we talk about sources of infection-where it starts, and the way or ways it spreads-the mode or modes of transmission. In the case of MRSA, patients who already have an MRSA infection or who carry the bacteria on their bodies but do not have symptoms (colonized) are the most common sources of transmission.

The main mode of transmission to other patients is through human hands, especially healthcare workers' hands. Hands may become contaminated with MRSA bacteria by contact with infected or colonized patients. If appropriate hand hygiene such as washing with soap and water or using an alcohol-based hand sanitizer is not performed, the bacteria can be spread when the healthcare worker touches other patients.

MRSA: a Growing Problem in the Healthcare Setting, but one with a Cure

MRSA is becoming more prevalent in healthcare settings. According to CDC data, the proportion of infections that are antimicrobial resistant has been growing. In 1974, MRSA infections accounted for two percent of the total number of staph infections; in 1995 it was 22%; in 2004 it was some 63%.

The good news is that MRSA is preventable. The first step to prevent MRSA, is to prevent healthcare infections in general. Infection control guidelines produced by CDC and the Healthcare Infection Control and Prevention Advisory Committee (HICPAC) are central to the prevention and control of healthcare infections and ultimately, MRSA in healthcare settings.

CDC welcomes the increased attention and dialogue on the important problem of MRSA in healthcare. CDC, state and local health departments and partners nationwide are collaborating to prevent MRSA infections in healthcare settings. For example, CDC:

- monitors trends in infections and MRSA through surveillance systems such as the National Healthcare Safety Network, formerly the National Nosocomial Infection Surveillance System and the Dialysis Surveillance Network to identify which patients are at highest risk and where prevention efforts should be targeted.
- works with multiple prevention partners including state health departments, academic medical centers, and

regional and national collaboratives to identify and promote effective strategies to prevent MRSA transmission.

- developed an overarching strategy to help guide healthcare facilities to control antibiotic resistance called The Campaign to Prevent Antimicrobial Resistance in Healthcare Settings. This campaign includes specific strategies for various healthcare populations, including hospitalized adults and children, dialysis patients, surgical patients, and long-term care patients.

An alarming number of microorganisms that occur in hospitals have become resistant against virtually all antibiotics. Several strains of pathogenic bacteria have been discovered that are resistant to essentially all known antibiotics. Resistance against so many different antibiotics has been induced by the indiscriminate use of heavy doses of antibiotics in health care. Strains that have developed resistance survive and can propagate productively as all competitors have been killed off. The discovery of bacterial strains with so many resistances has put development of new antibiotics into high gear.

One needs to have effective back-up antibiotics before bacterial strains with resistance to all currently known antibiotics have developed. The search for new antibiotics in many cases involves the combination of brute-force screening and targeted genetic modification of microorganisms that produce antibiotics: by selected modifications antibiotics with slightly different structure wand properties may be produced by these microbes. Nonetheless, the search for new antibiotics probably will need to be a continuing one. For example, anew type of antibiotic (Zyvox) was introduced in 2000, but the first Zyvox-resistant strains already were found a year later.

Many known antibiotics are produced by actinomycetes (particularly by representatives of the genus Streptomyces), which are filamentous soil bacteria. The antibiotics are produced via secondary metabolic pathways, and originate from a small number of simple precursors, including amino acids, small fatty acids, sugars, and nucleic acids. Secondary metabolism is more specialized than primary metabolism and varies widely even among members of the same genera. Several new antibiotics have been found by targeted or random inactivation of particular genes, leading to utilization of alternate biochemical pathways. It is still pretty much a matter of just testing out which mutants may give rise to novel and active antibiotics, but progress is being made towards this semi-rational design of such antibiotics. However, it will be some time before such antibiotics will be used as first a lengthy testing and certification procedure will need to be followed for each compound.

MILK PRODUCTION

By means of expression of foreign genes, essentially any protein from any source can be produced by bacteria. Barring the need for glycosylation or other post-translational processes to yield an active enzyme, large amounts of eukaryotic (plant, human, or animal) protein can be made by bacteria as long as they have been provided by the right gene construct. This can have a wide variety of implications, some of which are controversial. For example, a growth hormone of cows, bovine somatotropin (BST; sold by Monsanto under the trade name Posilac), has been produced in large quantities by genetically engineered micro-organisms. When given to the cows in small amounts, milk production in dairy cows increases by 10-25%. This increase in milk production per cow follows recent milk production increases obtained by traditional breeding: since 1950, milk production per cow has more than doubled. However, as BST leads to increased milk production, it will drive prices lower, which may force small farmers out of business.

However, the process of increasing monopolization by bigger companies is nothing new, and the BST-induced increase in production is peanuts compared to breeding-induced increases over the last 40 years. The FDA (Food and Drug Administration) has found the BST to be safe for human consumption. However, public is generally perceived to be wary of "BST cows." Some dairy products are marked to not have been obtained from BST-treated cows, and several supermarket chains claim to not sell any milk from BST-treated cows. The outcry from some consumer and small-farmer-interest groups has caused the FDA to perform additional trials, which backed up the initial findings, and BST is deemed to be safe in the dairy industry. In some respects the outcry regarding the use of genetically engineered hormones in lifestock is less than logical, in that lifestock has been fed antibiotics and hormones for years (to reduce infections and to increase production) without a lot of opposition from the consumer.

Even though the general discussion and perception of the BST issue in most cases does not have a rational or logical foundation, it forces people to start thinking about the effects of biotechnology in their everyday lives. Also, the issue provides a challenge to well-informed individuals to lead a public discussion to critically evaluate how appropriate, ethical, and healthy current lifestock management procedures really are.

BIOREMEDIATION

Biotransformation of Toxic Wastes to Harmless Products

The rapid expansion and increasing sophistication of the chemical industries in the past century and particularly over the last thirty years has meant that there has been an increasing amount and complexity of toxic waste effluents.

At the same time, fortunately, regulatory authorities have been paying more attention to problems of contamination of the environment. Industrial companies are therefore becoming increasingly aware of the political, social, environmental and regulatory pressures to prevent escape of effluents into the environment. The occurrence of major incidents (such as the Exxon Valdez oil spill, the Union-Carbide (Dow) Bhopal disaster, large-scale contamination of the Rhine River, the progressive deterioration of the aquatic habitats and conifer forests in the Northeastern US, Canada, and parts of Europe, or the release of radioactive material in the Chernobyl accident, etc.) and the subsequent massive publicity due to the resulting environmental problems has highlighted the potential for imminent and long-term disasters in the public's conscience.

Even though policies and environmental efforts should continue to be directed towards applying pressure to industry to reduce toxic waste production, biotechnology presents opportunities to detoxify industrial effluents. Bacteria can be altered to produce certain enzymes that metabolize industrial waste components that are toxic to other life, and also new pathways can be designed for the biodegradation of various wastes. Since waste management itself is a well-established industry, genetics and enzymology can be simply "bolted-on" to existing engineering expertise.

Examination of effluents from the chemical and petrochemical industries shows that such effluents typically contain either one or a limited range of major toxic components. In some cases other considerations (such as aesthetic ones) can be important for removal of certain components (such as dyes). This means that in general one industry may apply one or a few genetically modified bacterial strains to get rid of its major toxic waste. However, it may be important to contain the "waste-eating" bacteria within the manufacturing plant, and not release these with the waste water. In such cases, filter installations will have to be built to separate the bacteria from the effluent.

Of course, the bioprocesses for treating toxic effluents must compete with existing methods in terms of efficiency and economy. However, the biotechnological solution to the problem requires only moderate capital investment, a low energy input, are environmentally safe, do not generate waste (hopefully), and are self-sustaining. Biotechnological methods of toxic waste treatment are likely to play an increasingly key role both as a displacement for existing disposal methods and for the detoxification of novel xenobiotic compounds. On the other hand, however, it is important to limit the generation of both hazardous and non-hazardous waste as much as possible, and utilize recycling methods wherever possible.

Over the last few years, a number of companies have been established already to develop and commercialize biodegradation technologies. Existence of such companies now has become economically justifiable, because of burgeoning costs of traditional treatment technologies, increasing public resistance to such traditional technologies (ranging from Love Canal to the ENSCO incinerator plans in Mobile AZ years ago), accompanied by increasingly stringent regulatory requirements. The interest of commercial businesses in utilizing microorganisms to detoxify effluents, soils, etc. is reflected in "bioremediation" having become a common buzzword in waste management. Companies specializing in bioremediation (or, as it was known several years ago, in biodegradation technologies) will need to develop a viable integration of microbiology and systems engineering. As an example of a bioremediation company, Envirogen (NJ) has developed recombinant PCB (polychlorinated biphenyl)-degrading microorganisms with improved stability and survivability in mixed populations of soil organisms. The same company also has developed a naturally occurring bacterium that degrades trichloroethylene (TCE) in the presence of toluene, a toxic organic solvent killing many other microorganisms. A large number of similar companies can be found using a web search engine and an appropriate keyword (such as "bioremediation").

Microorganisms have also been successfully applied during the removal of the Exxon Valdez oil spill. A number of microorganisms can utilize oil as a source of food, and many of them produce potent surface-active compounds that can emulsify oil in water and facilitate the removal of the oil. Unlike chemical surfactants, the microbial emulsifier is non-toxic and biodegradable. Also, "fertilizers" have been utilized to increase the growth rate of the indigenous population of bacteria that are able to degrade oil. Use of microbes for bioremediation is not limited to detoxification of organic compounds. In many cases, selected microbes can also reduce the toxic cations of heavy metals (such as selenium) to the much less toxic and much less soluble elemental form. Thus, bioremediation of surface water with significant contamination by heavy metals can now be attempted. As is apparent that, the US government (in particular the US Department of Energy; DOE) has a keen interest in bioremediation (for example, see the NABIR (Natural and Accelerated Bioremediation Research) Program. Part of this interest stems from the commitment to clean up heavily polluted sites (such as the Hanford site in Washington state) that once were nuclear weapons facilities and that contain large, buried metal buckets of radioactive waste that now are starting to leak.

MICROBIAL INSECTICIDES

Pesticidal Proteins

Insect infestation can cause a significant decrease in crop productivity. Such infestations usually were difficult to fight without the use of nasty chemicals with high toxicity to other life. However, more environmentally sound approaches are being developed using biotechnology. One approach involves the Bacillus thuringiensis toxin protein. This protein all by itself is harmless, but is converted to a potent toxin in the gut of certain kinds of moths (depending on the bacterial strain the toxin was isolated from) and of mosquito larvae. The gene coding for this protein has been cloned from various Bacillus thuringiensis strains, and has been incorporated into several plants.

The moth against which the toxin is active dies after eating from the transformed plant. The drawback of this case of biological pest control is that the plant is resistant to that particular moth (even though the beast has first eaten from it before it is killed), but not to most other species of moths. Another drawback is that resistance of the moth to the toxin may develop (the only thing the moth has to do is to learn to not convert the original toxin protein in its midgut). The first cases of resistance to Bacillus thuringiensis toxin in caterpillars already have appeared. To minimize this problem, farmers now plant some "wild type" crop (not producing the toxin) in a corner cf their plot, and the toxin-producing crop of the same species on the remainder. In this way, toxin-resistant insects may not have such a huge advantage over their toxin-sensitive siblings that the toxin-resistant ones quickly become the prevalent species in the ecosystem.

The major advantage of the Bacillus thuringiensis (Bt) toxin is that it is harmful to only a few species of insects, while it is essentially harmless to other animals and humans. These biological pesticides also degrade rapidly in the environment. Thus, the use of such biological pesticides appears to be a significantly more environmentally safe solution to pest control than the classical (synthetic chemical) pesticides. Indeed, the majority of the remaining cotton fields in Arizona have been planted with transgenic cotton plants producing a Bt toxin that is particularly effective against the pink bollworm, the primary pest on cotton in Arizona. At least 4% of the fields planted with transgenic cotton is set aside to be planted with the non-transformed strain. This is part of the strategy to minimize the ecological survival advantage of Bt-resistant bollworms that may develop or that may "immigrate." Cotton seeds carrying a Bt gene have been commercially available since 1996.

Other Ways of Biocontrol

Another approach towards biological pest control is based on an original and rather devious idea: as many male insects are attracted to females through chemicals (pheromones) the females excrete in minute quantities, one can spray the fields with pheromones, thus profoundly confusing the males about where to find their partner. The genes for pheromone biosynthesis have been cloned from various insects and expressed in bacteria, thus paving the way for making enough pheromones to spray the fields with. Note, by the way, that pheromones by-and-large are innocuous compounds and need to be sprayed only in minute concentrations. Preliminary evidence indicates that this approach is highly successful.

Other molecular-genetically based techniques of environmentally responsible pest control are: (1) Sterilize a large number of male insects (for example, by irradiation), and release them in the field. They will mate, but no progeny will result. (2) Clone genes for the synthesis of juvenile or anti-juvenile hormones from insects, produce the hormones in large quantities, and spray on the fields. An excess of juvenile hormone will prevent maturation of the insects, and an excess of anti-juvenile hormone will result in premature maturation and sterility. These hormones generally are specific for certain groups of insects and are not toxic to others, thus minimizing the impact on the environment. However, presently biological pesticides still have a rather modest market share compared to the total pesticides marketed. The answer to this apparent paradox is that the narrow spectrum of control makes biological pesticides unattractive for some applications. However, the low market share is also due in part to the fact that the development of biological pesticides is relatively new (of the last dozen years), and that industry has been slow with catching on to the idea, and has not yet spent considerable resources on development of better biological pesticides. However, the public opinion currently is strongly in favor of ecologically acceptable methods of insect control, and this will impact the setting of priorities in product development by industry.

8

METHODOLOGY OF PLANT GENETIC ENGINEERING

PLANT TRANSFORMATION

Genetic engineering of plants is much easier than that of animals. There are several reasons for this:

(1) there is a natural transformation system for plants (the bacterium Agrobacterium tumefaciens),

(2) plant tissue can redifferentiate (a transformed piece of leaf may be regenerated to a whole plant), and

(3) plant transformation and regeneration are relatively easy for a variety of plants.

The soil bacterium Agrobacterium tumefaciens ("tumefaciens" meaning tumor-making) can infect wounded plant tissue, transferring a large plasmid, the Ti plasmid, to the plant cell. Part of the Ti (tumor-inducing) plasmid apparently randomly integrates into the chromosome of the plant. The integrated part of the plasmid contains genes for the synthesis of (1) food for the bacterium, and (2) plant hormones. Genes from the Ti plasmid that are integrated in the plant chromosome are expressed at high levels in the plant. Overproduction of the plant hormones leads to continuous growth of the transformed cells, causing plant tumors. Rapid, cancerous growth of the transformed plant tissue obviously is advantageous to the bacterium: more food gets produced. The Ti plasmid has been genetically modified ("disarmed") by deleting the genes involved in the production of bacterial food and of plant hormones, and inserting a gene that can be used as a selectable marker.

Selectable marker genes generally are coding for proteins involved in breakdown of antibiotics, such as kanamycin. Any gene of interest can be inserted into the Ti plasmid as well. In principle, one can thus transform any plant tissue, and select transformants by screening for antibiotic resistance. However, unfortunately, there are some complications:

(1) it has proven difficult to transform some monocots (grasses, etc.) by Agrobacterium, and

(2) regeneration of plants from tissue culture or leaf discs is not always possible.

A number of genetically engineered plant varieties have been developed. Traits that have been introduced by transformation include herbicide resistance, increased virus tolerance, or decreased sensitivity to insect or pathogen attack. Traditionally, most of such genetically engineered plants were tobacco, petunia, or similar species with a relatively limited agricultural application. However, during the past decade it now has become possible to transform major staples such as corn and rice and to regenerate them to a fertile plant. Increasingly, the transformation procedures used do not depend on Agrobacterium tumefaciens. Instead, DNA can be delivered into the cells by small, μm-sized tungsten or gold bullets coated with the DNA.

The bullets are fired from a device that works similar to a shotgun. The modernized device uses a sudden change in pressure of He gas to propel the particles, but the principle of "shooting" the DNA into the cell remains the same. This DNA-delivery device is nicknamed "gene gun", and has been shown to work for DNA delivery into chloroplasts as well. Over the last several years, use of the "gene gun" has become a very common method to transform plants, and has been shown to be applicable to virtually all species investigated. For example, transformation of rice by this method is now routine. This is a very important development as rice is the most important crop in the world in terms of the number of people critically dependent on it for a major part of their diet.

Another method to get foreign genes into cereals is by electroporation: a jolt of electricity is used to puncture self-repairing holes in protoplasts (i.e., the cell without the cell wall), and DNA can get in through these holes. However, it is often very difficult to regenerate fertile plants from protoplasts of cereals. Nonetheless, significant advances in overcoming these practical difficulties have been made over the years. Now even transgenic trees have been created: for example, the gene for a coat protein of the plum pox virus has been introduced into apricot. The plum pox virus leads to the feared Sharka disease, for which there is no cure. The resulting transgenic tree shows a markedly decreased sensitivity to this virus. The reason why continuous exposure of the tree to the viral coat protein leads to tolerance against viral infection is not yet understood, however.

Thus, now there are a number of different techniques to introduce foreign genes into plants. Essentially all major crop plants can be (and have been or are being) genetically engineered, the procedures are now routine and the frequency of success is very high. Even though genetically engineered crops are more costly than the usual ones, they have been rather readily accepted by US farmers provided that tangible benefits can be demonstrated. However, it is questionable whether the farmer in poorer countries can come up with the funds to "try out" and use the new crops. Another issue in this respect is how genetically engineered crops are perceived by the consumer.

Even though in the US there is little resistance to such crops as long as the products can be shown to be safe and advantageous, in other countries (for example in sections of Europe) genetically modified foods are received poorly by the consumer. It is unlikely that there is a rationally sound basis for this rather hostile reaction of the consumer, as most of the crops are the result of human manipulation (such as centuries of breeding) and may have been treated with harmful herbicides and pesticides. Time and education will need to be invested to provide consumers and consumer advocates with a balanced opinion on the acceptability of the origin of their foods.

FACTS ABOUT GENETICALLY ENGINEERED FOOD

Do you know that giant agrochemical companies have launched a massive venture to genetically restructure the world's food supply? Do you know...

- that the genetic blueprints of fruits, grains, and vegetables are being restructured with conglomerations of DNA from viruses, bacteria, insects, and animals?
- that over 60% of processed foods consumed in the U.S. now contain ingredients from genetically engineered crops?
- that numerous distinguished scientists view this genetic restructuring of food as a threat to both human and environmental health?
- that the U.S. Food and Drug Administration covered up the extensive warnings of its own scientific experts about the unique risks of genetically engineered foods and claims it is not aware of any information showing they differ from other foods?
- that besides violating sound science, the genetic restructuring of our food affronts the core principles of most religions?

Purpose and Goals

The Alliance for Bio-Integrity is a nonprofit, nonpolitical organization dedicated to the advancement of human and environmental health through sustainable and safe technologies. To this end, it aims (a) to inform the public about technologies and practices that negatively impact on health and the environment and (b) to inspire broad-based, responsible action that helps correct the problems and uphold the integrity of the natural order. In approaching these issues, it integrates the perspectives of both science and religion and coordinates the participation of both communities.

The Alliance's initial project is to gain a more rational and prudent policy on genetically engineered foods. This entails (a) educating the public about the unprecedented dangers to the environment and human health posed by the massive enterprise to genetically reprogram the world's food supply; (b) securing a scientifically sound system for safety-testing genetically altered foods; and (c) securing a meaningful system of labeling in order to protect the right of consumers to avoid such foods.

Achieving the latter two objectives requires an action at law, since current U.S. Food and Drug Administration policy exempts genetically altered foods from the testing required of new food additives and also permits these foods to be marketed without identifying labels. Although respected groups such as Consumers Union, the Union of Concerned Scientists, and the Environmental Defense Fund have strongly criticized this policy as scientifically flawed and unsound in several other respects as well, the FDA staunchly refuses to revise it.

Accordingly, the Alliance has organized an unprecedented plaintiff group to bring a lawsuit against the FDA to effect the necessary changes. The plaintiffs include eminent scientists, public interest organizations, and people from diverse faiths who reject genetically altered foods on the basis of religious principle. The suit was filed May 27, 1998 in U.S. District Court in Washington, D.C. and is being managed by the legal department of the International Center for Technology Assessment in Washington, which shares the Alliance's concerns about genetic engineering and has an impressive record in public interest litigation.

One area of particular concern for some people is the lack of labeling of genetically engineered foods, and legislation may be introduced to address this issue.

You are Eating Genetically Engineered Food

Is it good for you?

Do you have a choice?

Genetic engineering is the largest food experiment in the history of the world. We are all the guinea pigs. There are about 40 varieties of genetically engineered crop approved for marketing in the U.S. As a result, 60-70% of the foods on your grocery shelves contain genetically engineered (GE) components.

Genetically engineered foods contain substances that have never been a part of the human food supply. They are not subjected to rigorous pre-market safety testing. And THEY ARE NOT LABELED.

Is genetic engineering safe for you and your family? Safe for the environment? Safe for the future of mankind? No long-term studies have been done. No one can answer these questions.

On the other hand, as so many plants (soybean, corn, etc.) are genetically modified and the nature of the genetic modification is not necessarily easy to explain, it may be simpler to label those foods that are guaranteed free of "genetically modified organisms" or their products. However, keep in mind that essentially all agricultural products have been genetically modified by traditional breeding, so it may be difficult to define what is actually free of genetically modified organisms.

Several plant biotechnology companies have increased their efforts to provide information regarding the full, global scope of impacts of plant biotechnology.

DUPONT BIOTECHNOLOGY

Biotechnology holds a great deal of promise to enhance our lives and planet. With a world population expected to reach nearly nine billion by 2050, biotechnology offers new potential for sustainable living, healthy eating and battling diseases while reducing our footprint on the planet. DuPont is committed to sustainable growth through the application of science to create value for society and our shareholders.

We believe the broad field of biotechnology presents important opportunities that should be explored and developed to identify those safe and commercially viable applications that bring significant benefits to society. These opportunities arise in areas including food, materials, energy generation, polymers, sensors and electronics. Benefits may include lower cost, higher quality products and reduced reliance on fossil fuels along with other environmental benefits.

The DuPont Difference

For 200 years, DuPont has worked diligently to bring to life the miracles of science. Our approach-innovation, stewardship and sustainability-is grounded in our heritage in and commitment to thoughtful, careful science. Since our beginnings in refining and manufacturing gun powder, safety and the environment have been priorities. Best expressed by Charles O. Holliday, Jr., DuPont chairman and CEO: "Safety is the essence of DuPont. Unfortunately, many in industry have been reluctant to address concerns about the risks of biotechnology.

But we have to listen to the people who are now raising alarms. We don't have all the answers and to pretend we do, or to brush off concern as unfounded, is to be arrogant and reckless. Listening implies engagment and respect, and that requires initiative, patience and the willingness to build relationships that will provide a point of view and perspective that may be counter to our own."

New Knowledge, New Ways of Living

As a science company dedicated to thorough, forward-thinking research and development, DuPont delivers science-based solutions that make a difference in people's lives in food and nutrition; health care; apparel; home and construction; electronics; and transportation. Nylon®, Teflon®, Kevlar®, Nomex®-through years of research, we have conceptualized, researched, refined and created solutions that offer new choices and set new standards.

Du Pont Researches through the Years

1908: First investigation of synthetic fibers launched. Laboratory equipment at the company's first Experimental Station (circa 1903) featured bar slabs and gas lighting, the simplest kind of apparatus. Yet, it was the most up-to-date of its day.

1927: Fundamental research program established the foundation for many products; start of the 20th century "materials revolution." Neoprene, first commercially successful synthetic rubber and fruit of fundamental research, shown by Wallace Carothers.

1938: Development of nylon, the basis of polymer science and the beginning of wash-and-wear fabrics. Nylon was first spun mechanically on this crude machine. For a spinneret, researchers used a drug store hypodermic needle.

1940s: Teflon® first marketed after World War II, paving the road to its eventual use in billions of heavy-duty, non-stick pots and pans.

1958: Invention of Lycra®, first intended as a replacement for rubber, known for its ability to stretch up to six times its original length and now used in stretch clothing.

1960s: DuPont introduces Kevlar®-which is a material in everything from bullet-resistant clothing to bicycle tires, and brake parts to athletic shoes-and Nomex®-which are both used to make the clothes that firefighters and race car drivers alike wear for their unsurpassed resistance to heat.

1970s: Commercial production of Tyvek® from DuPont takes off, distinguished by its protection capabilities with wide use ranging from construction material to mailing envelopes, and automobile covers and medical packaging.

1990s: Bicyclers in the Tour de France ride to victory with DuPont CoolMax®, the fabric that eliminates moisture from the body, keeping the body cool and comfortable.

These company sites probably are just as subjective as some of the sites listed above that are very critical of genetically modified organisms, and the best solution is to read information provided by both sides and to see what is reasonable.

PLANT GENETIC ENGINEERING: APPLICATIONS

"Pharming" and "Plantibodies"

An increasingly viable option is the production of highly valuable enzymes by plants and animals. In addition to production of human proteins in these organisms, other valuable proteins that are currently produced by microorganisms could very well be produced by higher organisms instead. Animal and plant "bioreactors" in some respects may be superior to recombinant bacterial systems, because eukaryotes glycosylate proteins. Whereas the glycosylation pattern may be species-specific, appropriate glycosylation is often required for protein function. Production through these organismal systems may also be cheaper than cell fermentation techniques.

Two examples of production of human proteins in plants include the production of human serum albumin in transgenic tobacco and potato, and production of human insulin by tobacco. In both cases, the produced protein appears to be fully effective in humans. Unfortunately, however, one cannot raise his/her insulin level by eating transgenic tobacco leaves, as the protein in most cases will be broken down to amino acids before it reaches the blood stream. Therefore, in these cases one cannot escape the practice of protein isolation and purification before transgenic leaves are converted into drugs. Also antibodies are being produced in plants. Initially, the antibody's light and heavy chains were produced in different plants. But a subsequent cross of these two varieties resulted in progeny carrying assembled and functional antibodies.

Human Antibodies Produced by Field Crops Enter Clinical Trials

Down a country road in southern Wisconsin lies a cornfield with ears of gold. The kernels growing on these few acres could be worth millions--not to grocers or ranchers but to drug companies. This corn is no Silver Queen, bred for sweetness, but a strain genetically engineered by Agracetus in Middleton, Wis., to secrete human antibodies. This autumn a pharmaceutical partner of Agracetus's plans to begin injecting cancer patients with doses of up to 250 milligrams of antibodies purified from mutant corn seeds. If the treatment works as intended, the antibodies will stick to tumor cells and deliver radioisotopes to kill them.

Using antibodies as drugs is not new, but manufacturing them in plants is, and the technique could be a real boon to the many biotechnology firms that have spent years and hundreds of millions of dollars trying to bring these promising medicines to market. So far most have failed, for two reasons.

First, many early antibody drugs either did not work or provoked severe allergic reactions. They were not human but mouse antibodies produced in vats of cloned mouse cells. In recent years, geneticists have bred cell lines that churn out antibodies that are mostly or completely human.These chimeras seem to work better:this past July one made by IDEC Pharmaceuticals passed scientific review by the Food and Drug

Administration: The compound, a treatment for non-Hodgkin's lymphoma, will be only the third therapeutic antibody to go on sale in the U.S.

The new drug may be effective, but it will not be cheap; cost is the second barrier these medicines face. Cloned animal cells make inefficient factories: 10,000 liters of them eke out only a kilogram or two of usable antibodies. So some antibody therapies, which typically require a gram or more of drug for each patient, may cost more than insurance companies will cover. Low yields also raise the expense and risk of developing antibody drugs.

This, Agracetus scientist Vikram M. Paradkar says, is where "plantibodies" come in. By transplanting a human gene into corn reproductive cells and adding other DNA that cranks up the cells' production of the foreign protein, Agracetus has created a strain that it claims yields about 1.5 kilograms of pharmaceutical-quality antibodies per acre of corn. "We could grow enough antibodies to supply the entire U.S. market for our cancer drug--tens of thousands of patients--on just 30 acres," Paradkar predicts. The development process takes about a year longer in plants than in mammal cells, he concedes. "But start-up costs are far lower, and in full-scale production we can make proteins for orders of magnitude less cost," he adds.

Plantibodies might reduce another risk as well. The billions of cells in fermentation tanks can catch human diseases; plants don't. So although Agracetus must ensure that its plantibodies are free from pesticides and other kinds of contaminants, it can forgo expensive screening for viruses and bacterial toxins. Corn is not the only crop that can mimic human cells. Agracetus is also cultivating soybeans that contain human antibodies against herpes simplex virus 2, a culprit in venereal disease, in the hope of producing a drug cheap enough to add to contraceptives.The web-shy Planet Biotechnology in Mountain View, Calif., is testing an anti-tooth-decay mouthwash made with antibodies extracted from transgenic tobacco plants. CropTech in Blacksburg, Va., has modified tobacco to manufacture an enzyme called glucocerebrosidase in its leaves. People with Gaucher's disease pay up to $160,000 a year for a supply of this crucial protein, which their bodies cannot make.

"It's rather astounding how accurately transgenic plants can translate the subtle signals that control human protein processing," says CropTech founder Carole L. Cramer. But, she cautions, there are important differences as well. Human cells adorn some antibodies with special carbohydrate molecules. Plant cells can stick the wrong carbohydrates onto a human antibody. If that happens, says Douglas A. Russell, a molecular biologist at Agracetus, the maladjusted antibodies cannot stimulate the body into producing its own immune response, and they are rapidly filtered from the bloodstream. Until that discrepancy is solved, Russell says, Agracetus will focus on plantibodies that don't need the carbohydrates. Next spring the company's clinical trial results may reveal other differences as well.

THERAPEUTIC AND DIAGNOSTIC PLANTIBODIES

At a conference in York last week, delegates from Europe, USA and New Zealand learned of new developments in the use of plants as providers of both natural and genetically modified products for the pharmaceutical and healthcare industries

Plants are already a valuable source of lead compounds for the pharmaceutical industry. According to Melanie O'Neill of Glaxo Wellcome's Medicines Research Centre, 8 of the top 30 medicines are natural products or semisynthetics with a value of $15.9bn in 1999. However, there is a continuing drive to discover new medicines for diseases that are poorly treatable and compounds with novel mechanisms of action.

The seminar learned how plants could provide such new lead compounds, new plant extracts and new combinations of genes, and also provide a delivery mechanism for therapeutic recombinant proteins and vaccines with benefits for developing countries. Neil Robinson of MolecularNature illustrated the rich diversity of the British flora in providing new leads for the pharmaceutical, healthcare and agrochemical industries. Based at IGER (Institute for Grassland and Environmental Research) in Aberystwyth, MolecularNature is developing advanced chemical fingerprinting techniques for the isolation of rare natural compounds with biologically relevant structures. Of the 250,000 plant species that exist on our planet, only about 38,000 have been studied phytochemically and, perhaps, only 2% of plants have been thoroughly evaluated as a source of new medicines. Aspirin from willow, Digitalis from foxglove, for the treatment of cardiac conditions, and the anti-cancer activity of Taxol from yew are the tip of the iceberg.

At the molecular level, John Bedbrook of Maxygen, USA explained how the application of his companies MolecularBreedingTM technique of 'DNA shuffling' could produce novel compounds in plants. This technique has yielded enzymes with improved kinetics, altered substrate specificities, altered optima and selectivity of optima. Also at the molecular level, but with a global perspective, Julian Ma of Guy's Hospital described how using plants as the expression system for antibodies, antigens and immune complexes could make vaccines more widely affordable in developing countries. 'Plants are the most efficient producers of protein on the planet, with simple nutritional requirements and the potential to be grown on an agricultural scale'. As a dentist, his work is currently focused on the development of monoclonal antibodies in topical passive immunotherapy applications. However, the technology has far reaching potential.

Organised by ACTIN (Alternative Crops Technology Interaction Network) and the University's Plant Protein Club, the seminar may have been the first time that organisations interested in developing recombinant and endogenous plant products have met under the same roof! However, Ian Bartle, ACTIN Chief Executive, expressed the wish that the event would lead to a longer term dialogue between people who have one thing in common, namely the belief that plants are a safe and cost effective delivery system for many high value products for today's market.

REVERSIBLE MALE STERILITY IN PLANTS

As has been indicated earlier, heterozygous individuals often are healthier and stronger than homozygous ones. The only way to guarantee heterozygoiscity in plants is to make sure self-pollination cannot occur. For most crop plants it was very tedious or practically impossible to exclude selfing. To exclude self-pollination, it would be good to introduce male sterility in plants: progeny from such plants are then expected to be 100% heterozygous (assuming they were pollinated with pollen from an unrelated variety).

To introduce male sterility, a promoter was identified that was turned on exclusively in tapetum cells (a tissue around the pollen sac that is essential for pollen production). This promoter then was linked up to a gene coding for a bacterial ribonuclease (named barnase). This ribonuclease selectively chops up ribonucleic acids. The promoter/ribonuclease construct was then introduced into plants (canola, tobacco, you name it). Because the promoter allows expression only in tapetum cells, the gene construct disrupts only development of the tapetal tissue and its end product, pollen. Plants transformed with this construct were male-sterile but otherwise normal. Although male-sterile plants are valuable for hybrid seed production, they have limited value when it comes to crop production. Fertility must be restored to crops such as wheat, rice, and tomato, in which the seed or fruit is the harvested product. Fortunately, the ribonuclease is inhibited very much by a simple protein, named barstar. One can thus cross the male-sterile plant with a male-fertile variety in which the gene for barstar has been introduced, and the result is progeny with viable pollen and restored fertility. A closely related approach has been criticized as "terminator technology" as it is seen by its critics as a way for companies to protect and enforce their patents. In any case, several genetically modified crops with barstar and barnase are available.

Antisense RNA

Antisense RNA refers to nucleotide strands that are produced in a cell and that are complementary to a particular mRNA. Antisense RNA can be produced, for example, by inverting the coding region of a gene with respect to its promoter. The antisense RNA can hybridize with its corresponding mRNA, making it double-stranded. The double-stranded mRNA no longer can be recognized by the protein-synthesizing machinery (the ribosomes), and thus expression of this mRNA is suppressed. Also, in many systems double-stranded mRNA is very unstable and is broken down quickly. Thus, one can inactivate specific genes while not interfering with others. Antisense approaches already are used to protect plants from damage by plant viruses. For example, reversal of a gene from bean yellow mosaic virus (BYMV), and putting it into tobacco under a reasonably strong promoter, has led to a tobacco variety that is quite resistant to BYMV.

Potyvirus Resistance Derived from Antisense RNA and Native or Chimeric Coat Protein Genes

An antisense (AS) RNA construct consisting of the C-terminal portion of the coat protein (CP) gene and complete 3' non-coding sequence of bean yellow mosaic virus (BYMV), and driven by the cauliflower mosaic virus (CaMV) 35S promoter, was used to obtain transgenic Nicotiana benthamiana plants by Agrobacterium-mediated transformation. Other plants were transformed with constructs designed to express the BYMV CP gene or chimeric CP genes. The original transformants from each construct were allowed to self. R1 plants carrying the introduced gene were selected on the basis of polymerase chain reaction (PCR) and/or ELISA (for CP-expressing plants) with monoclonal antibodies. Homozygous R2 plants expressing AS RNA displayed a range of resistance, from minimal to apparent immunity from infection by BYMV; no resistance was observed to other potyviruses.

Plants expressing native BYMV CP also showed a range of resistance, with a minimal degree of resistance to other potyviruses. Both AS and CP plants displayed two types of resistance; to initial infection, and/or to replication or movement. Chimeric CPs, with the N-terminal domains of BYMV fused to the C-terminal domains of pepper mottle potyvirus or zucchini yellow mosaic potyvirus, differed in their response to challenge with several potyviruses. At least one transformant of each chimeric CP showed milder symptoms than non-transgenic controls when inoculated with BYMV, and some resistance to potato virus Y. Deleted constructs are being prepared with the aims of separating the two types of resistance, determining the mechanisms of resistance, and which domains confer viral specificity. A similar approach is used to transfer viral resistance to other plants. This finding is of significance, in that currently no effective, environmentally friendly methods exist to control many plant viruses.

Very related to this approach is the RNAi (RNA interference) approach. This is very useful for both agriculture and medicine, and the first examples of practical applications of this RNAi technology are appearing. As with any new technology, the initial pilot projects are sort of pedantic (including manipulation of flower color and of the speed of fruit ripening). However, more exciting application possibilities abound. Obviously, RNAi technology provides an excellent approach for reverse genetics in eukaryotes. With this method one can turn off genes, and see what the consequences are.

Agricultural Applications in Developing Countries

Perhaps indicative of the large potential and relative ease of genetic engineering, developing countries (particularly China) are progressing rapidly in development and application of genetically engineered crops. Some have gone into commercial production well ahead of similar crops in the US.

In China, for example, tomatoes that have been engineered for improved virus resistance have been on the market since late 1992. There are two main reasons for the more rapid commercialization of bioengineered crops in the developing world: (1) less tight governmental approval mechanisms, and (2) hungrier populations. While in developed countries the main value of biotechnological applications may be to reduce production costs, in the developing world a main factor is the production of more food. Indeed, genetic engineering applications seem to be pretty successful to cut down on pathogen-induced losses. For example, genetic modification of papaya plants (expression of the ringspot virus coat protein in the plant) protects very well against the very destructive ringspot virus.

TRANSGENIC ANIMALS

In some cases overexpression of human genes in bacteria (such as E. coli) does not yield a protein that is functionally active in humans. The reason for this is that some proteins need to be post-translationally modified (phosphorylated, glycosylated, etc.) before they are active. Bacteria generally lack the specific enzymes recognizing the human protein sequences that need to be modified, and thus the bacterially produced gene product will differ from the native one. To counter this problem, certain human genes can be introduced into farm animals (usually yeast will do the job, too), and when these genes are expressed in the mammary glands of the animals, the post-translationally modified protein can be isolated from milk, tested whether its post-translationally modified product is identical or at least very similar to the native human one, and if so, be developed as a pharmaceutical.

For example, the genes for two different human blood clotting factors (VIII and IX) have been hooked up to sheep and pig regulatory sequences that causes expression in mammary tissue; after transformation of sheep or pig embryos, genetically engineered animals have been selected that produce milk with a large percentage of human blood-clotting factor. This protein can be isolated from the milk, purified, and marketed. Similarly, transgenic rabbits have been created that produce human interleukin-2, which is a protein stimulating the proliferation of T-lymphocytes; the latter play an important role in fighting selected cancers.

Other human proteins that have been expressed in transgenic animals include: anti-thrombin III (to treat intravascular coagulation), collagen (to treat burns and bone fractures), fibrinogen (used for burns and after surgery), human fertility hormones, human hemoglobin, human serum albumin (for surgery, trauma, and burns), lactoferrin (found in mother milk), tissue plasminogen activator, and particular monoclonal antibodies (including one that is effective against a particular colon cancer). Animals mostly used for this work are pigs, cows, sheep, and goats.

The amounts of milk needed to provide a national supply of these pharmaceuticals are really very reasonable. Assuming the animals produce 1 g of the protein per liter milk and one has a purification efficiency of 30% (that is, 30% of the protein is recovered in the pure sample), then a pig can produce 75 g of protein per year, a goat 100 g, a sheep 125 g, and a cow 3 kg. As the national need of blood-clotting factor IX is 2 kg/yr, one cow per country can do the job. For other proteins the demand is larger (for example, for tissue plasminogen activator it is 75 kg per year and for human serum albumin it is about 1,000 kg/yr), but nonetheless a limited number of animals is all one now needs to meet the national demand for pharmaceutical proteins that used to be astronomically expensive.

DOLLY AND POLLY

A fairly large stir was caused in the popular media when a group in Scotland associated with the Roslin Institute and with PPL Therapeutics announced in early 1997 that a lamb, Dolly, had been born that had been cloned from a single cell taken from her mother's udder. Of course, cell division of vegetative cells to eventually yield a new organism with the same genetic makeup as the parent is pretty usual among "lower" organisms and certain plants, but until 1997 it had never been shown for mammals. The creators of Dolly had taken an unfertilized egg cell with the nucleus removed, and fused that with the cell from the udder. The fused cell was made to divide and developed into a normal embryo. This was implanted into a surrogate mother, and it developed into a healthy lamb. This was the first time that genetic information from a fully differentiated, vegetative mammalian cell was used to give rise to a new, fully differentiated organism.

Building upon the "success" of Dolly, the next step came the same year from the same group in Scotland. The single, diploid cell originating from the adult sheep now was genetically altered (introducing a human gene gene coding for blood clotting factor IX) before fusing with a denucleated egg cell. The fused cell was made to divide and to develop into an embryo, which was implanted into a surrogate mother. The resulting lamb, Polly, contains the human gene in every cell of her body.

The method resulting in Polly is seen as a major improvement as compared to the technology of the early nineties that led to the first transgenic bovine creature, Herman, carrying the human lactoferrin gene (lactoferrin is an important component in breast milk). At that time, genes were injected into newly fertilized eggs, and only in rather infrequent cases did the gene stably integrate and lead to a transformed animal.

Now that Dolly is an adult, questions are being raised regarding her health. In most respects she is a normal sheep. She is fertile and has given birth to a number of lambs. However, at a young age she developed arthritis, which usually does not occur in sheep until much later. Whether or not this is a consequence of cloning (which may not set back the biological clock) is as yet unknown. Her telomers are a little shorter than usual for an animal of her age, but whether this has an impact on her health and longevity is unclear.

Dolly in most respects was a normal sheep. She was fertile and has given birth to a number of lambs. However, at a young age she developed arthritis, which usually does not occur in sheep until much later, and she died at 6 years of age (much younger than the average lifetime of sheep) due to a progressive lung disease. Whether or not this is a consequence of cloning (which may not set back the biological clock) is as yet unknown. Her telomers were shorter than usual for an animal of her age, but whether this had an impact on her health and longevity is still unclear.

Many more animal clones have been generated in the mean time. For example, cloned cows appeared in 1999 and now there are cloned pigs that have been modified to reduce transplant rejection of pig organs in humans. Cloned pets (cats and dogs) have been created too. There are even cloned mules. The success with mammalian cloning has led to a large flood of responses, most of which are related to ethical issues regarding mammalian cloning. Indeed, what can be carried out with most mammals can also be carried out with humans, and therefore some very valid ethical issues on where to draw the line of the acceptable can be raised. An additional issue that has received a lot of attention is the use of human embryonic stem cell lines for medical purposes. This will be covered later.

FISH FARMING

In some countries transgenic fish has been developed. This is quite easy, as there is usually no problem to get female gametes: just squeeze the female; no surgery, no microscopes. The egg cells can be just electroporated to introduce the desired DNA. However, one does not need "high tech" approaches to increase aquaculture yields. Often it is sufficient to have fish male and female hormones produced in bacteria, and utilize these hormones. For some fish, after hatching a male fingerling exposed to estrogen will become female in appearance, while remaining genetically male. The "pseudofemale" can lay eggs producing viable offspring, in spite of the male chromosomes. Female fingerlings can be sex-reversed in the same way through exposure to testosterone, becoming reproductively viable pseudomales.

This application is attractive to fish farmers, who often want all-male groups of fingerlings: they grow faster than females and single-sex ponds mean that a second generation of fingerlings ("recruits") is not produced when the stocked fish becomes sexually mature. Recruits will eat, but will not be marketable by harvest time, thus bringing down production.

However, understandably consumers are not very enthused about having fish that has been exposed to hormones. To avoid having to hormone-treat fingerlings that will be harvested later, one can use a male parent with 2Y chromosomes and no X. All progeny will be XY and male, without needing any hormone treatment. YY males are indistinguishable from XY males and can be obtained from a normal male (XY) and a pseudofemale (a male that was treated with estrogen). Half of the resulting progeny is normal XY male, 1/4 is female (XX) and 1/4 is YY ("supermale"). XYs and YYs can be distinguished from each other by DNA typing. If one wants to produce progeny that is 100% female, one can simply cross a "pseudomale" (XX) with a normal female.

HUMAN MOLECULAR GENETICS

Sequencing Genomes

One of the major areas of progress during the past decade has been the generation of an enormous amount of DNA sequence generation. For an ever increasing number of organisms a complete genome sequence is now known. The sequence of several bacteria, including Haemophilus influenzae, Mycoplasma genitalium, the archaebacterium Methanococcus jannaschii, and the cyanobacterium Synechocystis sp. PCC 6803 has been completed in the mid nineties. The Escherichia coli genome was completed in 1997. Since then the sequence of hundreds of other prokaryotes has been completed or is in progress.

The Institute for Genomic Research (TIGR) in Maryland and the Kazusa DNA Research Institute in Japan played a major role in early sequencing efforts. Yeast has the distinction of being the first eukaryote of which the complete genome sequence has been finished (in 1996). A large number of microbial and other genome sequences are now generated at the DOE-funded Joint Genome Institute in Walnut Creek CA. With improved techniques and increasing capacity and accuracy of automated sequencers, bacterial genomes now can be sequenced within a few days if one has a sufficiently large number of automated sequencers lined up. However, any of these sequencing projects are about three orders of magnitude smaller than the Human Genome Project.

The first eukaryote with a larger genome that had its genome sequence completed was a nematode (worm) by the name of Caenorhabditis elegans (C. elegans for short). The nematode genome is about 97 million base pairs long, contains about 20,000 genes, and sequencing was essentially completed in 1998. Groups involved in the sequencing effort were at the Sanger Center near Cambridge (UK) and at Washington University (St Louis, MO), and therefore this sequencing effort was done in an academic (rather than industrial) setting. As a demonstration project to illustrate the feasibility of their sequencing approach for complex genomes, the Perkin-Elmer/ Venter initiative (in collaboration with scientists at UC-Berkeley) sequenced the genome of Drosophila, the fruitfly.

The Drosophila genome is about 4% of the size of the human one. The project uses 230 automated sequencing machines working in parallel to collectively churn out 100 million base pairs per day. The initiative started in April 1999, and in August 2000 most of the 120 million nucleotides long Drosophila sequence had been determined. However, the sequence is still being annotated and remaining gaps in the sequence are still being closed several years later. This illustrates the difficulty to generate a full genomic sequence for eukaryotes. In addition to determination of the human genome sequence, genomes of other mammals (mouse, rat) are also being sequenced. The first plant with a virtually complete genome sequence is Arabidopsis thaliana (about 125 million nucleotides).

The Arabidopsis Information Resource (TAIR) maintains a database of genetic and molecular biology data for the model higher plant Arabidopsis thaliana. Data available from TAIR includes the complete genome sequence along with gene structure, gene product information, metabolism, gene expression, DNA and seed stocks, genome maps, genetic and physical markers, publications, and information about the Arabidopsis research community. Gene product function data is updated every two weeks from the latest published research literature and community data submissions. Gene structures are updated 1-2 times per year using computational and manual methods as well as community submissions of new and updated genes. TAIR also provides extensive linkouts from our data pages to other Arabidopsis resources.

The first agronomically important plant with a sequenced genome is rice (about 400 million nucleotides); a draft sequence of the genome of two rice varieties was published in 2002, and a more complete sequence came out in 2005. The first fish with a completed genome sequence was the puffer fish (2002), with the zebra fish second.

The results of the smaller sequencing efforts have made it clear that there is much to be learned from a chromosome or genome sequence. Very importantly and unexpectedly, a large number of long open reading frames identified in the genome sequences does not have known homologues with similar sequence in other organisms. This is new information, and the function of the products of these putative genes remains to be established. In the case of yeast and Synechocystis sp. PCC 6803, this can be done relatively easily by making targeted mutations in these putative genes, and analyzing the effect of these introduced mutations. For other organisms, less direct but nonetheless fairly effective methods (for example, RNAi) to "knock out" genes have been developed. This will provide information on what type of protein that particular gene is coding for. In the case of humans, however, the introduction of targeted mutations is not desirable.

THE HUMAN GENOME PROJECT

The entire human genome is about 3 billion nucleotides long. In the mid-nineties, sequencing cost less than $1 per nucleotide (it currently is much cheaper than that), and the idea came up to find funds to sequence the entire human genome.

Certainly an ambitious project, but why not? The project, called the human genome project (HGP), had an unusual origin. It was not initiated by a committee of molecular geneticists, or by the major biomedical funding agency, NIH. Instead, it was proposed by an administrator in the Department of Energy (DOE), convinced that the powerful tools of molecular biology made it appropriate to introduce centrally administered "big science" into biomedical research. Later on, the National Institute of Health (NIH) jumped on the HGP bandwagon as well. Before long the Human Genome Project had become an international venture, with participation by Japan, England, Italy, and other countries in the Human Genome Organization (HUGO). HUGO serves as an international umbrella for information exchange and collaborations. The Human Genome Project was started up with James Watson (from the Nobel-prize winning Watson & Crick double-helix) as director. Dr. Watson resigned in 1992, but his initial leadership provided the program with sufficient impetus that it had gone beyond the "point-of-no-return". The project was planned to take 15 years (completion in 2005), and initially progress was slow. The initial phase was to provide a physical and linkage map of all human chromosomes, providing the genetic location of diseases with a genetic basis on one of the chromosomes.

For this purpose, thousands of "probes" (relatively small pieces of expressed DNA of 1,000-5,000 nucleotides in length, and with a unique sequence) that are spaced at regular intervals of about 100,000 nucleotides were identified. This information already had a medical payoff, in that genetic screening for a particular disease became much easier. After making a genetic map of all chromosomes, brute-force sequencing was applied and all nucleotide sequences were put into huge databases in which sequences were put together and in which introns, open reading frames, sequence repeats, etc. were directly analyzed.

The publicly funded Human Genome Project, which aimed at a complete sequence by 2005 and which appeared behind schedule, received a significant boost (or kick in the behind?) from a private effort spearheaded by Perkin-Elmer Inc. (the main DNA sequencer manufacturer) and J. Craig Venter, the former president and director of The Institute for Genomic Research (TIGR) who subsequently headed the Perkin Elmer sequencing subsidiary, Celera Genomics. This private effort aimed at sequencing the entire human genome by 2001 at a cost of $150-300 million using a shotgun cloning and sequencing approach they pioneered for bacterial genomes.

The publicly funded Human Genome Project was supposed to cost 10 times more. The bottom line is that the Human Genome Project and Celera Genomics made a joint announcement in June 2000 that a working draft of the DNA sequence of euchromatin regions (containing most genes) of the human genome has been developed. At that time, about 65% of the genome had been sequenced... well ahead of the initial schedule. The remainder of the sequence has now been determined and assembled, except for sequence repeat regions and other regions that are hard to interpret and assemble. Attention has now turned to finding where genes lie in the DNA sequence and what functions those genes control. This information can be used to fashion better medical treatments and to shape the direction of development for new drugs.

A main difference between publicly and privately funded genome projects may be timely release of sequence data. Perkin-Elmer/Celera can sell access to annotated databases as well as patent rights of genes to pharmaceutical companies and biotech firms. However, raw sequence data are made available for free.

Initially, the human genome was thought to contain up to 100,000 genes, occupying about 3% of the total genome. The function of much of the remainder of the genetic material is as yet unclear. With the majority of the human genome sequence in hand, it is clear that the number of genes has been overestimated by about a factor of 3, and that humans (with their 30,000 genes or so) have just 2-fold more genes than a simple plant, worm, or fruitfly, and just 10-fold more genes than most bacteria.

The human genome sequence, written down as a "word" of over 3 billion letters, fills the equivalent of 134 sets of the complete volumes of the Encyclopaedia Brittanica.

A company as well as a federal government agency (NIH) have attempted to patent gene sequences of unknown function; this was not accepted by the patent offices to be patentable. However, even though random sequences are not patentable, it is possible that complete genes can be patented for direct use (for example, for a testkit). But even if patenting is not straightforward, companies where new sequence is generated (or who contract with sequencing companies), will be able to capitalize on this information before anyone else learns of its existence. Therefore, the most justifiable and ethically correct attitude in this respect may be to have all sequence information enter the public domain expediently, and commercial enterprises who wish to capitalize on this information may do so if desired. This will avoid secrecy in data gathering (which would lead to unnecessary duplication of efforts), and maximize openness.

When sequencing large stretches of DNA with hitherto unknown sequence, it is of utmost importance to be highly accurate in the sequence analysis. Omission of a single nucleotide in a sequenced gene will shift the reading frame, and will result in an entirely erroneous derived amino acid sequence behind the location of the sequence analysis error. Currently, a 99.97% accuracy is quoted in sequence determinations. Even though this suggests that DNA sequencing is extremely precise (which it is), it should be kept in mind that there may be 3 errors in every stretch of 10,000 nucleotides of determined sequence.

The main source of errors is the omission or addition of a single nucleotide, and therefore there may be a reading-frame shift every 3,000 nucleotides or so (on average). Keeping in mind that genes generally are somewhere between 100 and 10,000 nucleotides in length, it is likely that a significant percentage of gene sequences that have been determined carry a frame shift. Therefore, effort is being made to further improve the DNA sequencing accuracy in genome projects. However, with the genome sequence being determined for a large number of different organisms, it becomes much easier to spot regions that are expected to contain frame shifts (how??).

USE OF GENOMIC DATA

The amount of information that results from genomic sequencing projects is stunning and overwhelming. In the case of eukaryotes containing introns in their genes, a first challenge is to accurately assign introns so that coding sequences can be correctly predicted. In most organisms this is not trivial. Part of the assignment depends on monitoring codon usage: most organisms have some codons that are not or rarely used, and for the third nucleotide of a codon (remember that there is sequence degeneracy at mainly the third position for many codons!) generally there is a clear preference depending on the organism. Prokaryotic assignments usually are simpler because of the lack of introns and because of the smaller genome. However, the start codon for the start of translation in prokaryotes is not always AUG (as it is in eukaryotes) but can also be GUG, CUG, or UUG, and, at lower frequency, selected other codons. Therefore, the search for start codons is a little more complex than in eukaryotes.

Once a genomic sequence has been determined and open reading frames (possibly translated regions, from start codon to stop codon) have been defined, then one can try to assign a function to the proteins coded for by the various open reading frames. The assignment of a function often involves comparing the sequence with sequences of known proteins from other organisms. A useful software program in this respect is BLAST. However, there are many open reading frames that do not have known homologues in other organisms, and in such cases one will need to resort to experiments such as deletion mutagenesis of the open reading frame in an appropriate organism, etc.

Once a genome sequence is known, it becomes possible to do genomic expression analysis, and determine which genes are expressed under which conditions. This information may also provide clues regarding the function of open reading frames coding for unknown proteins. Methods have been developed to obtain comprehensive insight in the expression of most genes in an organism. These methods generally utilize a DNA chip (also known as gene chip or microarray), in which a collection of hundreds or thousands of genes (cDNAs) or gene fragments (oligonucleotides of 25-70 bases long) from one organism has been spotted on a microscope slide. After isolating mRNA from this same organism (perhaps from a particular tissue or when grown under specific conditions), one can do a reverse transcription of this mRNA to complementary DNA (cDNA), incorporating a fluorescent label in the cDNA. The cDNA is then hybridized to the blot or microarray, and the relative level of expression for each gene can be determined in a single experiment. The drawback of this genomic analysis using microarrays is that this requires expensive chemicals and equipment.

Gene chips provide information about transcript levels. While this information is important and informative, transcript levels do not necessarily correspond to the amount of protein coded for by these transcripts. As protein ultimately is the material that is functionally relevant (enzyme activity, structural function, etc.), there is much interest in developing ways to determine the level of many different proteins in a tissue or cell. Antibody arrays have been developed, in which a large number of different antibodies have been attached to different locations in an array. After extraction of proteins from a tissue and labeling them with a fluorescent marker, the mixture can be incubated with an antibody array, and the relative intensity of the spots on the array (each representing the relative amount of crossreaction with a specific antibody) can be determined.

GENETIC DISORDERS

A number of diseases are linked to a defect in a specific gene. In such cases, it is important to find the faulty gene: then treatment may become easier and more direct (for example, by regular injections with the gene product in the case of insufficient gene expression or of a mutation within the coding region, as now is done to treat diabetes). Mapping of a certain disease to a defect in a specific region of the genome used to be a huge job. For example, to map and sequence the gene for cystic fibrosis took six years of research by a large number of groups, at an expense of approximately \$150 million (5% of the cost of the entire human genome project). With a genomic sequence in hand, a genetic disease is mapped, narrowing down the disease locus to 10-100 genes. Based on predicted gene function, the locus may be narrowed down further, and candidate genes may be sequenced from patients and healthy individuals. Gene localization may now be accomplished in a matter of months, at a fraction of the cost of gene localization in the pre-genomic era.

In many cases, genetic disorders are due to single nucleotide mutations. However, in some cases small genome rearrangements may have taken place, affecting the expression of specific genes. These rearrangements may not have led to changes in gene sequence. However, they can be readily determined by Southern blotting. Total genomic DNA (obtained from readily available cells, such as white blood cells) is chopped up using restriction endonucleases (recognizing specific 4-8 bp sequences in the DNA) into fragments of variable length that are separated by electrophoresis. The DNA fragments are transferred to nylon membrane, and probed with the gene or

DNA region of interest. If there were major rearrangements or sequence alterations in the region of the gene in the patient, then the length of the fragment(s) obtained from patients with the disease may differ from the length obtained from unaffected individuals.

To test for single-base mutations, the simplest method is SNP (single nucleotide polymorphism) analysis using a microarray carrying oligonucleotides with similar but non-identical sequences. Comparing relative hybridization intensities, the sequence at specific loci can be predicted reliably without the need for large-scale sequencing. Identified regions may be sequenced to confirm the conclusion from SNP array analysis.

The reason for the large interest of companies in SNP analysis is obvious: A collection of thousands of relatively frequently occurring SNPs linked to genetic disorders may be put on a single DNA chip, and people may be screened for any mutations that may potentially impact their health. For a healthy individual, the presence of mutations that are linked to an increased occurrence of a particular disease may lead to increased vigilance in this respect, whereas for a patient the reason for particular symptoms may be diagnosed expediently.

Restriction Fragment Length Polymorphism (RFLP)

The genome sequence of each individual is different. Also, within one individual the two homologous chromosomes (one originating from each parent) have differences. On the average, 1 out of 100-500 nucleotides is different between two individuals. Of course, the number of differences is smaller when individuals are closely related. In many cases, these genetic variations between individuals are neutral: one variant is neither more helpful nor more harmful for survival and reproduction than the other. These variations usually occur in regions of the genome that do not encode for structural genes. They are more likely to be found in introns, pseudogenes, and sequences between genes of no known function than in sequences that are known to encode proteins. The genotypic variations that do not have phenotypic effects are called silent mutations.

The inter-and intraindividual sequence variations often are concentrated in some regions of the genome, and are less frequent in others. Thus, in certain genome regions sequences are variable, and virtually any individual may have their own, unique pair of DNA sequences for this region. This property is taken advantage of by the technique of restriction fragment length polymorphism (RFLP) mapping. A specific sequence may contain a restriction site for a certain enzyme in one of the chromosomes of one individual, but the site may be absent in another person.

Thus, when using a probe of cloned DNA around this sequence, the restriction fragment lengths from genomic DNA (on a Southern blot) hybridizing to the probe may be different in different persons. Using different restriction enzymes, individuals that have a similar pattern with one restriction enzyme may have different patterns with another enzyme. Thus, it is possible to obtain a "DNA fingerprint" that is different for each individual, provided that sufficient probes for different polymorphic DNA regions are used, and that a number of different restriction enzymes have been applied.

In addition to obvious advantages of RFLP mapping for forensic applications, RFLPs are extremely useful as markers to localize disease-related genes on specific chromosomes. For example, if in one family a certain disease is found to be linked to the occurrence of a specific restriction fragment length in a variable region in different siblings, whereas other family members who do not have the disease do not show that RFLP pattern, it is likely that the RFLP locus and the locus where the disease-causing allele is located are linked. Thus, if it is known where that particular RFLP region is, then it is known where (approximately) to find the mutated gene. A RFLP probe repository has been established to facilitate the exchange of probes between investigators, so that it becomes possible to establish relationships between markers. One of the first diseases localized by genetic linkage analysis using RFLPs was Huntington's disease.

This disease is genetically dominant, shows up at 35-40 years of age, and causes a progressive degeneration of the central nervous system. Death follows within several years after the first symptoms occur. As this disease is dominant, it is simple to follow. For recessive disorders, the use of RFLP to locate genes associated with the genetic disorder is less trivial, since it may not be known until several generations later who is a carrier of the disease (heterozygote without disease symptoms) and who is not. To map a recessive disease to a specific chromosome and location, it often is necessary to use families in which at least two siblings with the disease and healthy parents are available for study.

All affected siblings within one family will have the same pair of disease-causing alleles, and will share RFLP patterns for probes linked to the gene associated with the disease. Healthy siblings and parents will not have the same RFLP pattern as siblings afflicted with the disease. This approach has been successful in locating the cystic fibrosis locus to the long arm of chromosome 7, and in eventually cloning and sequencing the gene that is altered in cystic fibrosis patients. The genetic defects that are the causes for retinoblastoma and for sickle-cell anemia (the latter being a mutation in the globin gene) have been mapped by this methodology to chromosomes 13 and 11, respectively.

The sequence of the human genome is providing us with the first holistic view of our genetic heritage. While not yet complete, continued refinement of the data bring us ever closer to a complete human genome reference sequence. This will be a fundamental resource in future biomedical research.

The 46 human chromosomes (22 pairs of autosomal chromosomes and 2 sex chromosomes) between them house almost 3 billion base pairs of DNA that contains about 30,000-40,000 protein-coding genes. The coding regions make up less than 5% of the genome (the function of the remaining DNA is not clear) and some chromosomes have a higher density of genes than others.

Most of the genetic disorders featured here are the direct result of a mutation in one gene. However, one of the most difficult problems ahead is to find out how genes contribute to diseases that have a complex pattern of inheritance, such as in the cases of diabetes, asthma, cancer and mental illness. In all these cases, no one gene has the yes/no power to say whether a person has a disease or not. It is likely that more than one mutation is required before the disease is manifest, and a number of genes may each make a subtle contribution to a person's susceptibility to a disease; genes may also affect how a person reacts to environmental factors. Unraveling these networks of events will undoubtedly be a challenge for some time to come, and will be amply assisted by the availability of the sequence of the human genome.

A complication (but at the same time a helpful feature) of genome mapping is recombination between chromosomes during meiosis. This phenomenon is known as crossing over. If the RFLP region and the locus of the genetic defect are close together, the probability of crossing over in between the two regions is low. However, if the distance is larger, the probability of recombination in between the two regions becomes larger as well. A longer distance (and thus more frequent crossing over between the RFLP marker and the gene of interest) complicates the interpretation of linkage studies. However, on the positive side, one can also calculate the approximate distance between the marker and the gene from the frequency that a certain RFLP pattern does not co-inherit with a certain disease. Identifying a gene once the locus has been "mapped" is not trivial, but the availability of the genome sequence is a great help. For example, a linkage between a certain RFLP pattern and a certain allele that has a chance of more than 95% to remain linked upon meiosis merely means that the allele and the RFLP region are not more than some 5 million nucleotides from each other.

With a high map density of RFLPs a still closer marker may be found, but as the markers get closer to each other, recombination between them will be observed so rarely that it will be impossible to determine with a reasonable degree of confidence where the gene of interest is located with respect to the RFLP markers. However, on average there are not more than 10-100 genes in a 5 million nucleotide span on the human genome, and when a disease has been mapped to such a region, one can first study the predicted function of the genes in this region and see whether their impairment might lead to symptoms found in the disease. If so, such genes (and their surrounding regions) may be analyzed first. If no obvious candidate genes are identified, then expression of genes in the mapped region may be monitored, and the reason for possibly aberrant expression patterns for a specific gene may be identified. In a number of cases a disease has been found to be linked to the sex of the individual. For example, hemophilia, red-green colorblindness, and Duchenne Muscular Dystrophy almost exclusively occur in males. If such a strong sex-linkage is found, it is evidence that the gene related to the disease is on the X chromosome: as males have only one X chromosome, X-linked traits that are recessive in females are essentially dominant in males. This sex linkage in some cases has been known for a long time. A good example in this respect is hemophilia, a deficiency in clotting factor VIII, which impairs healing of wounds: the Talmud, written about 1,500 years ago summarizing oral laws of Jewish religion, tells of a Rabbi instructing a woman not to have her son circumcised after three of her sister's sons had bled to death. But no exception would be granted if her brother's sons (rather than her sister's) had met a similar fate. Explain why.

It should be kept in mind that a number of diseases are multi-factorial: they are caused by a combination of factors, none of which by itself would cause the disease. This complicates matters greatly, in that it is very difficult to track such a disease genetically. On the other hand, diseases with very similar manifestations may be caused by mutations in different genes. This again complicates identification of genetic changes correlated with such a disease. Explain why.

A map has been compiled of the human chromosomes with about 11,000 approximate location(s) of mutations or loci linked to genetic diseases.

GENETIC SCREENING

Genetic screening for common diseases already is a common part of obstetrical care and a standard procedure for newborns. Also, in cases where there are major medical concerns regarding the health of an unborn baby on the basis of genetic diseases occurring in the family of either parent, specific tests can be carried out on the genetic information carried by the foetus. In many cases, cells from the amniotic fluid that bathes the foetus are used as the test material for either biochemical or genetic screens. Similar information also can be obtained by chorionic villus sampling. In this procedure, the doctor removes a piece of tissue from the developing placenta for study: this procedure can be carried out earlier in pregnancy. Also preconception counseling and screening can be carried out: for example, in some communities of Hasidic Jews, couples are screened before marriage to detect whether both carry one defect gene for hexosaminidase A. If so, there is a 25% chance that a child will not be able to produce this enzyme, and will suffer of Tay-Sachs disease. Children who inherit two impaired hexosaminidase genes suffer severe neurological effects and typically die before age four.

The occurrence of relatively rare recessive diseases may be much more frequent in communities that traditionally have been "self-supporting" in terms of finding a partner in life. Inbreeding significantly increases the probability with which recessive diseases surface. In this respect Tay-Sachs disease already has been mentioned, which is much more frequent in Hasidic Jews than in other populations. Another example is the Ellis-Van Creveld syndrome ("six-fingered dwarfism"), of which 50 cases are known among the Amish population in Lancaster County PA. Worldwide, only another 50 cases of this disease have been recorded. A little closer to home is relatively high occurrence (0.5%) of albinism among the Hopi Indians. This albinism, due to tyrosinase deficiency that causes a lack of melanin (pigment) synthesis, occurs at a 0.002% rate worldwide.

Routine genetic screening can be as simple as counting chromosomes (an extra chromosome 21 is indicative of Down's syndrome). If specific genetic diseases run in a family, then genetic screening can involve RFLP probing related to genes that are associated with the disease (thalassemia, sickle-cell anemia, hemophilia, or cystic fibrosis, to name a few). More than 4,000 inherited diseases are due to single-gene defects. The RFLP restriction pattern of the foetus (or young person) being screened is then compared to that of various family members, some with the disease and some without. From this comparison, it is usually possible to determine whether the foetus or young person will develop the disease or not. This information is advantageous for early diagnosis of a disease.

In many cases an effective treatment of the disease is possible if diagnosed at an early stage, often even earlier than when the first symptoms develop. A good example for this is phenylketonuria (PKU), which is due to a lack of phenylalanine hydroxylase, which converts the amino acid phenylalanine to the amino acid tyrosine. Absence of a functional enzyme leads to accumulation of phenylalanine and toxic derivatives. The latter result in seizures, mental retardation, and a decreased life span. Blood of infants usually is tested for unusually high levels of phenylalanine. If a high phenylalanine level is found, this may indicate phenylketonuria. The PKU symptoms and damage can be avoided by having a phenylalanine-free diet, particularly during the first 15 years of life. PKU patients should also avoid aspartame (NutraSweet), as this is a phenylalanine derivative.

Results of simple genetic and/or biochemical screening also can provide information regarding someone's ancestry. Most examples to date relate to biochemical blood analysis. For example, essentially all American Indians carry the so-called Rhesus factor (which protein got its name by reacting with an antiserum from Rhesus monkeys). Virtually the entire indigenous population of East Asia does so too, thus confirming the hypothesis of common ancestry of the two groups. Absence of the Rhesus factor is quite common among blacks and whites. Tribes of Native Americans, such as the Sioux, that have mixed more with whites and blacks have a larger proportion of rhesus-negative individuals (not carrying the rhesus factor) than tribes that have remained relatively "pure" for a longer time (such as the Navajo). As another example, whites can gauge their heritage by their blood group: blood group B is thought to have entered Europe with the invasion of the Tatars from Asia. Thus, if someone has blood group B (or AB), this most likely means the individual has some Tatar ancestry. The reverse conclusion (blood group O or A, thus no Tatar ancestry) is not justified; explain why. Similar population-genetic tests can be done by RFLP mapping.

Most intriguing, and potentially most troubling, are genetic tests that peer well into the future. Huntington's disease, for example, typically appears after age 30; a person with an affected parent stands a 50% chance of developing the fatal disorder. Being tested for the single deadly gene can bring a powerful surge of relief, or the certainty of progressive mental and physical decline (no treatment is yet available). For some inherited adult-onset disorders, however, advance warning can serve to influence the course of the disease. Adult polycystic kidney disease invariably leads to kidney failure, but detecting concomitant high blood pressure early and effectively controlling it may delay the need for chronic kidney dialysis. Tests for susceptibility to certain malignancies, such as retinoblastoma (an eye cancer already occurring during infancy and childhood), or familial intestinal polyposis (which usually progresses to colon cancer) permit stepped-up vigilance and early treatment. Also, in 1993 two genes (msh2 and mlh1) have been identified that predispose people to non-polyposis colon cancer; this type of cancer strikes one in 20 people, and in about 20% of those cases the cancer is linked to specific mutations in the msh2 or mlh1 genes. About 10 companies already have purchased the rights to develop msh2 and mlh1 tests, and thus have staked their claim in this huge market. Another gene, brca1, has been linked to breast-and ovarian cancer. For this another presymptomatic gene test with a huge market has been developed. However, a number of important questions should be asked regarding ethical implications of these gene tests: Is it ethical to test for diseases for which there are no known cures? How reliable are the available tests? What are the psychological consequences for healthy patients of learning their possible destiny? Is the regulation of laboratories that offer genetic testing stringent enough to ensure that life-shattering errors are not made? How can perfectly healthy people who may carry a defective gene be protected from discrimination by health and life insurance companies and potential employers?

One key issue in this respect is whether the knowledge provided by gene testing will actually save lives. For Huntington's disease, the answer is clearly no as no cure is available. For cancers, the answer is not clear. In general, early detection of cancers is associated with improved survival. But the question is whether interventions that work for the general population are adequate for individuals with a strong genetic risk. For example, mammograms that are used to detect breast cancer early may not be good for people with brca1 mutations, as low doses of radiation might conceivably trigger further mutations that could lead to cancer. The approval mechanism for new genetic testing materials involves the Food and Drug

Administration. They were specifically charged with this task through a recommendation by the Secretary's Advisory Committee on Genetic Testing (currently subsumed in the Secretary's Advisory Committee on Genetics, Health and Society) at NIH. This website has interesting documents on genetic discrimination. Three percent of the Human Genome Project budget is set aside to address ethical considerations. In particular, use of genetic screening by insurance companies or prospective employers are areas of considerable concern.

FORENSIC APPLICATIONS

Each individual carries a unique set of genetic information. Also, the genome is different between parent and child, and between children of the same parents (with the exception of identical twins): the child is the product of half of the genetic information of both parents, and it is by chance which chromosome of each pair in each parent is transmitted to a certain child. Crossover during meiosis further complicates the inheritance pattern. In humans there are more than a million different ways to combine chromosomes from two parents, and crossover during meiosis adds a myriad of possibilities to recombine between two homologous chromosomes of each parent.

Since all individuals have their own unique set of DNA sequences, it is possible to identify everyone by their DNA. In the 1990's, RFLP mapping was used to compare the DNA from an individual with the DNA from a sample (for example, hair left at the site of a crime), "proving" or disproving that the DNA sample came from the individual. Disproving is simple (any change in the RFLP pattern is usually significant), but proving that a sample came from a certain individual is much more difficult: many different probes will need to be used to be statistically certain that this individual (and perhaps the identical-twin sibling) is the only one in the entire universe with all these RFLP patterns.

To simplify analyses in forensic laboratories, moderately repetitive DNA (some 10-50 copies, often scattered throughout the genome) was used as a probe. Because of the repetitive nature of this DNA, probing with such DNA gives a number of restriction fragment lengths, thus obtaining a large number of data with a single probe. However, it is possible that there are some sequence differences between the probe and some copies of the repetitive sequences to which it hybridizes. This generally results in some stronger and some weaker bands. Moreover, the region of homology between the probe and some DNA fragments in some cases is relatively small, causing the band on the Southern blot to be quite weak.

The different bands visualized on one autoradiogram offer an obvious advantage: using one probe and one blot, a whole lot of different bands can be found, each of which can be viewed as an independent "witness for the prosecution" or "witness for the defense". If each of the bands are absolutely identical between the DNA from the suspect and the DNA from the site of the crime, the probability can be estimated that this would be coincidence: this probability most likely is vanishingly small. However, if one or more bands do not correspond between the two samples, it is quite certain that the two samples do not come from the same individual.

A large problem regarding the use of gene technology for forensic purposes in many instances used to be the limited availability of DNA that is to be compared to that of the suspect. For a nice Southern blot, a few μg's of DNA are required. By comparison, one single diploid cell contains only 12 pg of DNA. So, about a million cells would be needed, provided that all DNA can be extracted efficiently.

Therefore, more modern forensic techniques make use of automated DNA amplification systems by means of the polymerase chain reaction (PCR), as was discussed in a previous section. In this way, it is possible to amplify small regions of the genome, starting from DNA extracted from one or several cells. If PCR primers are chosen to border regions with a variable number of tandem repeats (VNTR), the sizes of PCR products resulting from a particular set of primers will depend on the individual whose DNA is used for amplification. Therefore, this information can be used for DNA fingerprinting as well.

The theory of DNA fingerprinting is crisp and clear. However, there have been a few cases of where the experiments and interpretations of DNA fingerprinting were rather equivocal, and to avoid this high standards have been set. The central problem is that the "real world" of a blood-soaked crime scene is a far cry from the controlled and near-immaculate environment of a research lab where the technology is being applied. While DNA in the lab is pure, DNA from cells gathered at the scene of the crime may be degraded, or may be contaminated with DNA from cells from other individuals, including the victim. Often more bands show up on the autoradiogram of DNA from the crime scene than in control DNA from the suspect and the victim.

Another difficulty is determining how likely two DNA fingerprints are to match by chance. This probability very often is linked directly to the quality of the data. Nonetheless, DNA fingerprinting constitutes a very valuable addition to the forensic tests currently available. One of the companies who was a leader in the field of forensic DNA testing is Lifecodes (first taken over by Orchid, which also took over CellMark, another DNA testing company, and now by Tepnel). They also offer paternity testing, based on the same DNA fingerprinting principles. Sometimes, such testing can lead to surprising results: In one case it was found that the child of whom the father was in dispute was not the biological son of his mother. It was suspected that an inadvertant swap of kids had occurred in the hospital. In another case, upon in-vitro fertilization sperm vials appeared to have been swapped and a baby could be shown to have a biological father different from the spouse of his mother.

The history of DNA use for forensic cases already spans almost two decades. The first cases into which DNA evidence was brought in were in England. The first case of using DNA-related evidence in Arizona courts was the 1988 murder of Jennifer Wilson by Richard Bible near Flagstaff. Blood found on the back of Bible's plaid shirt was identified through DNA testing as Jennifer's blood--with a probability of 14 billion-to-1. Bible was subsequently convicted. This conviction was upheld unanimously by the Arizona Supreme Court. This together with other legal opinions elsewhere have paved the way for further use of DNA-related evidence in trials. National and international scrutiny of this method has occurred during the OJ Simpson trial in the mid-nineties. While many uncertainties may remain after the trial, the admissibility and validity of

DNA evidence in courts had clearly been established. Currently, DNA evidence is frequently used by the courts, and challenges of DNA-based evidence have virtually disappeared.

A rather unique (and sad) case of the utility of biotechnology in solving family issues was necessitated by action of the former military government in Argentina. About 12,000 people were taken as political prisoners and "disappeared" between 1976 and 1983. Children whose parents thus "disappeared" usually were adopted by military personnel, sold, or in some instances "disappeared" as well (a phenomenon still occurring on the streets of Rio de Janeiro). Grandmothers tried to keep track of what the fate of their grandchildren was. After a change to a slightly more democratic government in 1983, a number of grandmothers tried to reclaim their grandchildren.

But how to prove that someone indeed is your grandchild? Testing for different enzyme types and blood groups provided a 99.8% certainty of a relationship between grandparents and grandchild, but that was insufficient proof for the courts. Then a more reliable method was applied, essentially identical to that used for identification of the czarina and three of her children: hypervariable regions of the mitochondrial genome (that do not code for proteins) were sequenced. As mitochondria (the power plants of the cell) are inherited only via the mother, the sequence of the mitochondrial genome of a grandchild is identical to that of the maternal grandmother. By comparing the sequence of sufficient mitochondrial DNA, the relationship between grandmother and grandchild could be unequivocally established.

OTHER APPLICATIONS OF DNA FINGERPRINTING

Currently the principal applications of DNA fingerprinting (or DNA typing) are in forensic analysis, paternity disputes, and immigration cases. However, DNA marker systems have found many other actual or potential uses. These include: (1) monitoring the success of bone marrow transplants, (2) animal breeding, and (3) conservation biology.

Bone marrow transplants are used most frequently in people being treated for leukemia. In this operation, the patient's own cancerous bone marrow is destroyed by a combination of radiation and chemical therapy, and replaced by normal bone marrow from a healthy tissue-matched donor. Since blood cells are made in the bone marrow, the success of the operation can easily be monitored by blood DNA typing to check that the DNA in the circulating blood is that of the donor, not the patient. Any reappearance of the patient's own DNA in the blood might signal the reappearance of cancerous cells, and appropriate therapy can then be instituted.

Remarkably, the multilocus probe systems developed for human DNA typing also produce highly variable and informative patterns from a wide range of animals, birds, reptiles, amphibians and fish, and in some cases even from invertebrates. Single-locus minisatellite and microsatellite markers are also being developed from non-human species. DNA typing is already providing animal breeders with a powerful new tool. For example, stolen animals can be identified from their DNA fingerprints. Similarly, DNA typing can be used to verify the identity of semen used in artificial insemination programs, and also to establish the pedigree of the animal. Indeed, several cases involving disputes over whether a champion dog had really sired a given puppy have been satisfactorily resolved by DNA fingerprinting. In the long term, DNA markers will make it possible to construct genetic maps of domesticated animals and thereby enable the eventual localization of genes controlling economically important traits, such as disease resistance, milk yield, and body weight.

In terms of being a tool in conservation biology, DNA typing is already helping to protect endangered species in various ways. It provides for the first time a method of identifying animals stolen from the wild. This can be achieved either by setting up a database of DNA fingerprints of wild animals against which a captive individual can be compared, or by showing that young individuals held by a breeder could not be the offspring of any other individuals that the breeder has in stock. A second application in this respect is found in helping zoos in their breeding programs, in particular by identifying closely related individuals and thereby minimizing the risk of inbreeding. More generally, DNA typing is beginning to revolutionize our understanding of the genetic makeup and breeding systems of natural populations, knowledge of which is of fundamental importance in monitoring the genetic diversity and reproductive success of natural animal populations. In addition, in the presence of DNA fingerprints of species in protected areas, poached animals can be easily identified. Several public and private institutions have specialized in molecular-genetic analysis of wildlife.

What is Gene Testing? How does it Work?

Gene tests (also called DNA-based tests), the newest and most sophisticated of the techniques used to test for genetic disorders, involve direct examination of the DNA molecule itself. Other genetic tests include biochemical tests for such gene products as enzymes and other proteins and for microscopic examination of stained or fluorescent chromosomes. Genetic tests are used for several reasons, including:

- carrier screening, which involves identifying unaffected individuals who carry one copy of a gene for a disease that requires two copies for the disease to be expressed
- preimplantation genetic diagnosis (see the side bar, Screening Embryos for Disease)
- prenatal diagnostic testing
- newborn screening
- presymptomatic testing for predicting adult-onset disorders such as Huntington's disease
- presymptomatic testing for estimating the risk of developing adult-onset cancers and Alzheimer's disease
- confirmational diagnosis of a symptomatic individual
- forensic/identity testing

In gene tests, scientists scan a patient's DNA sample for mutated sequences. A DNA sample can be obtained from any tissue, including blood. For some types of gene tests, researchers design short pieces of DNA called probes, whose sequences are complementary to the mutated sequences. These probes will seek their complement among the three billion base pairs of an individual's genome. If the mutated sequence is present in the patient's genome, the probe will bind to it and flag the mutation. Another type of DNA testing involves comparing the sequence of DNA bases in a patient's gene to a normal version of the gene. Cost of testing can range from hundreds to thousands of dollars, depending on the sizes of the genes and the numbers of mutations tested.

GENES—DREAMS AND REALITY

Headlines about DNA, genes, and the new powers of scientists to analyze and manipulate these fundamental elements of life vie for our attention daily. The dazzling diversity of applications of DNA science to fields ranging from medicine and agriculture to forensics and environmental restoration undoubtedly will have resounding impacts on society and each of our own lives. Many of the new genetic discoveries stem from data and tools generated by the massive international Human Genome Project (HGP), whose goal is to describe in intricate detail the DNA from humans and other selected organisms by 2003. Because DNA is the information molecule that carries instructions for creating and maintaining all life, resources and analytical technologies generated by the HGP and other genetic research can be applied to the DNA of all organisms on earth. Other important HGP goals are to develop tools for data analysis and to address some of the ethical, legal, and social issues that may arise from the project. This article offers some basic information on the Human Genome Project and DNA science that will help the reader understand why genetic information is so powerful. It also describes a few current applications and some developments we can expect to see in the next few years, and presents some of the potentially troubling societal concerns surrounding this work.

The Human Genome Project

The HGP began in 1986 as a way for scientists in the U.S. Department of Energy (DOE) to use newly developing DNA analytical technologies to fulfill a long-standing mandate from Congress to assess the health effects of radiation. For decades DOE and its predecessors have developed international standards for the use of advanced medical diagnostic tools and treatments involving radiation and the protection of workers in the federal and civilian nuclear industry.

As the potential benefits of human genetics research became more apparent, Congress requested that DOE and the U. S.

National Institutes of Health develop a joint genome project. The U.S. Human Genome Project began formally in 1990 with expanded goals to describe all human genetic material (DNA) by 2005. However, rapid technological achievements advanced the expected completion date to 2003, and a draft product is eagerly anticipated by 2000. International research teams, particularly those from the United Kingdom but also from France, Germany, and Japan joined U.S. scientists to make significant contributions to the HGP. Today researchers worldwide are using HGP data and powerful analytical technologies to devise creative applications in an expanding array of fields.The claims and promises of these new capabilities are diverse and bold. But the new technologies and the data they generate also present complex ethical and policy issues that individuals and society, including the courts, have begun to confront (see "Societal concerns of the new genetics").

Societal Concerns of the "New Genetics"

Fairness in the use of genetic information by insurers, employers, courts, schools, adoption agencies, and the military, among others. Privacy and confidentiality of genetic information.

Psychological impact and stigmatization due to an individual's genetic differences. Reproductive issues including adequate informed consent for complex and potentially controversial procedures, use of genetic information in reproductive decision making, and reproductive rights.

Clinical issues including the education of doctors and other health service providers, patients, and the general public in genetic capabilities, scientific limitations, and social risks; and implementation of standards and quality-control measures in testing procedures. Uncertainties associated with gene tests for susceptibilities and complex conditions (e.g., heart disease) linked to multiple genes and gene-environment interactions.

Conceptual and philosophical implications regarding human responsibility, free will vs genetic determinism, and concepts of health and disease. Safety and environmental issues concerning genetically altered foods and microbes.

Commercialization of products including property rights (patents, copyrights, and trade secrets) and accessibility of data and materials.

DNA AND DISEASE

For all our apparent outward diversity, humans are surprisingly alike at the DNA level. We differ by only one or two tenths of one percent of our DNA-some three to six million bases-yet these tiny DNA variations are responsible for all our physical differences and may inßuence many of our behaviors as well. Most DNA variation among individuals is normal, but harmful variations called mutations can cause or contribute to many different diseases and conditions. Depending on their size and where in the DNA they occur, mutations can have devastating effects or none at all. If they occur within genes, the result can be the creation of faulty proteins that function at less-than-normal levels or are completely nonfunctional and result in disease.

All diseases have a genetic basis. We may inherit a particular condition, such as the lung disease cystic fibrosis, or an increased likelihood for developing such disorders as heart disease or colon cancer. We also inherit the particular ability to respond to such environmental stresses as viruses, bacteria, and toxins. Understanding how DNA inßuences every aspect of health eventually will lead to far more effective ways to treat, cure, or even prevent the thousands of diseases that afflict humankind. Some 4000 rare diseases are due to a single mutation in a single gene. These include cystic fibrosis, sickle cell anemia, and Tay Sachs. The causes are much more complex for common disorders such as heart disease, diabetes, hypertension, cancers, Alzheimer's disease, schizophrenia, and manic depression. These diseases are thought to be due to a variety of gene mutations, perhaps acting in concert, or to a combination of genes and such environmental factors as diet or exposure to radiation or toxins. Untangling the genetic and environmental contributions to complex disease will be one of the greatest challenges for medical researchers in the next century.

DNA Science Applied: Medicine and Health

Gene tests. DNA-based tests are among the first commercial applications of the new genetic discoveries to medicine. These tests are employed to diagnose a condition or estimate the likelihood for developing one. Test results already are being offered as evidence to support medical and nonmedical cases in courts, including medical malpractice, discrimination, privacy violations, child custody disputes, and criminal cases.

Gene tests involve direct examination of the DNA molecule itself. A DNA sample can be obtained from any tissue, including blood. To do a gene test, scientists scan the sample, looking for a specific mutation in a particular DNA region that has been linked to a disorder. Cost can range from hundreds to thousands of dollars, depending on the sizes of the genes examined and the number of mutations tested for, which can vary from a few to hundreds.

Although there are several hundred DNA-based tests for different conditions, most are still offered as research tools only. Fewer than 100 gene tests are available commercially, and most are for mutations associated with rare diseases in which just a single gene is involved.

SOME CURRENTLY AVAILABLE GENE TESTS

Gene tests for the disorders listed below are available from clinical genetics laboratories around the country. (Test name and some symptoms appear in parentheses.) The gene tests currently available (most in research settings only) detect only rare conditions that are usually caused by DNA changes in a single gene. Such common diseases as hypertension, heart disease, diabetes, and many cancers have complex genetics probably involving several genes that interact with environmental conditions to cause disease. There are no gene tests for these conditions yet, but this undoubtedly will change as more is learned about DNA.

- Amyotrophic lateral sclerosis (ALS; Lou Gehrig's Disease; progressive motor function loss leading to paralysis and death)
- Gaucher disease (GD; enlarged liver and spleen, bone degeneration)
- Inherited breast and ovarian cancer (BRCA1 and 2; early-onset tumors of breasts and ovaries)
- Hereditary nonpolyposis colon cancer (CA; early-onset tumors of colon and sometimes other organs)

- Cystic fibrosis (CF; disease of lung and pancreas resulting in thick mucous accumulations and chronic infections)
- Duchenne muscular dystrophy/Becker muscular dystrophy (DMD; severe to mild muscle wasting, deterioration, weakness)
- Fanconi anemia, group C (FA; anemia, leukemia, skeletal deformities)
- Fragile X syndrome (FRAX; leading cause of inherited mental retardation)
- Hemophilia A and B (HEMA and HEMB; bleeding disorders)
- Huntington disease (HD; usually midlife onset; progressive, lethal, degenerative neurological disease)
- Neurofibromatosis type 1 (NF1; multiple benign nervous system tumors that can be disfiguring; cancers)
- Adult Polycystic Kidney Disease (APKD; kidney failure and liver disease)
- Prader Willi/Angelman syndromes (PW/A; decreased motor skills, cognitive impairment, early death)
- Sickle cell disease (SS; blood cell disorder; chronic pain and infections)
- Spinocerebellar ataxia, type 1 (SCA1; involuntary muscle movements, reßex disorders, explosive speech)
- Thalassemias (THAL; anemias)
- Tay-Sachs Disease (TS; fatal neurological disease of early childhood; seizures, paralysis)

Even though some current gene tests have been beneficial and their potential benefit enormous, the science is very new and dynamic. Researchers themselves are unsure how to interpret the results of some commercially available gene tests (see "Gene tests: the power and the limits").

GENE TESTS: THE POWER AND THE LIMITS

Scanning a person's genes for mutations linked to a particular disease already has saved some lives and dramatically improved others. Some gene tests can alert patients and physicians to an inherited tendency toward a disorder and lead to increased surveillance or other preventive treatments. For example in familial adenomatous polyposis (FAP), a rare form of inherited colon cancer, lives have been saved through testing for the mutated gene linked to FAP and aggressive monitoring for early removal of colon growths or even the entire colon.

Interpreting the meaning of a negative test for the FAP mutation, however, is not straightforward. The possibility of developing the disorder is not ruled out because different mutations may be responsible for the disease in different individuals.

However, some physicians who order tests and interpret them for patients are unaware of these subtleties. An article published in the New England Journal of Medicine reported that one-third of physicians in a study group misinterpreted negative results in the FAP mutation test. If the researchers monitoring the study had not intervened, those doctors would have advised their patients to discontinue aggressive surveillance (colonoscopies), advice that could have had disastrous consequences. Comprehensive education of medical professionals is considered critical to the effective introduction of the new genetics into clinical practice.

Another limitation is the lack of medical options to treat or prevent many of the disorders for which gene tests are used. Researchers acknowledge the long lag time between linking a gene mutation with a disease and developing effective therapeutics.

Additionally, patients agreeing to undergo gene testing face significant risks of jeopardizing their employment and insurance status. Patients face an additional burden as well: the psychological impact of testing can be devastating. Because genetic information is shared, all these risks extend to family members as well. Many in the medical establishment feel that uncertainties surrounding test interpretation, the current lack of available medical options for most of these diseases, the potential for provoking anxiety, and the risks of discrimination and social stigmatization could outweigh the early benefits of testing.

Who's Regulating Gene Tests?

Most gene tests are offered as clinical laboratory services (rather than self contained "kits"), and while the U.S. Food and Drug Administration (FDA) has the authority to regulate such services, it has chosen not to because of a lack of resources. Although the quality of a laboratory to perform a test accurately is regulated under the Clinical Laboratory Improvement Amendments of 1988, no regulations exist that require evidence of a particular gene test's clinical validity (the probability that a person who tests positive will actually develop the disease) or it's utility (the potential for preventing or delaying the development of the disease in a person with a positive test).

People who are educated in these medical uncertainties are less likely to choose gene testing when they are weighing their benefits against the possibilities of discrimination by insurers, employers, schools, and others. Some companies have exaggerated both the validity and clinical utility of current gene tests in their eagerness to market these first commercial products of the "new genetics."

Although most current gene tests are used to diagnose or predict a risk for developing rare diseases, testing for susceptibility to more common diseases--like heart disease and diabetes--is the largest category of tests under commercial development. We can expect that aggressive marketing (some have called it "genohyping") will increase with the widening spectrum of tests developed, and some may in fact be offered directly to the public, a situation already occurring in the United Kingdom.

The Secretary's Advisory Committee on Genetic Testing of the U.S. Department of Health and Human Services is presently exploring these and other medical, scientific, ethical, legal, and social issues raised by the development and use of genetic tests. The committee has also sought public perspectives on these issues as it prepares its recommendations, which are due in the spring. Preventive medicine and customized therapies. Studies of gene function will lead to a deeper understanding of normal biological processes and how they go awry in disease states. These insights will allow the development of better and earlier predictive tests and eventually usher in a field of prevention-based medicine and diagnostics.

Within the next decade, researchers also will begin to understand how DNA variations underlie our individual responses to medical treatments. Tens of thousands of people are hospitalized each year as a result of toxic responses to medications that are beneficial to others.

Some cancers respond dramatically to current therapeutic regimens while the same treatment has no effect on disease progression in others. Scientists in major pharmaceutical companies are trying to sort out the specific regions of DNA associated with drug responses, identify particular subgroups of patients, and develop drugs customized for those populations. These capabilities are expected to make drug development faster, cheaper, and more effective while drastically reducing the number of adverse reactions.

Drug design itself will be revolutionized as researchers use gene sequence and protein structure information to create new classes of medicines based on a reasoned approach rather than the traditional trial-and-error methods for finding new drugs. The new drugs, targeted to specific sites in the body and to particular points in the cascade of biochemical events leading to disease, will likely cause fewer side effects than many current medicines. Ideally, they would act earlier in the disease process.

Gene therapy and genetic enhancement. The potential for using genes themselves to treat disease has captured the imagination of the public and the biomedical community. This rapidly developing field-called gene transfer or gene therapy-holds great potential for treating or even curing such genetic and acquired diseases as cancers and AIDS by using normal genes to replace or supplement defective genes or bolster a normal function like immunity.

Over 350 clinical gene-therapy trials are now in progress worldwide, most for different kinds of cancers. Performed on patients in advanced stages of disease, most current studies aim to establish the safety of gene-delivery procedures rather than determine their effectiveness. The technology itself still faces many obstacles before it can become a practical approach for treating disease; however, novel experimental approaches look very promising.

GENE THERAPY

One of the most intriguing applications of genetic research is the use of genes themselves to treat, cure, and ultimately prevent disease. The science of gene therapy is in its infancy, however, and the goal of most current clinical trials is only to demonstrate the procedure's safety, not its effectiveness. A partial listing follows of diseases that are the focus of clinical gene-therapy trials.

Canavan disease

Cystic fibrosis

Familial hypercholesterolemia

Gaucher's disease

Hemophilia B

Various advanced cancers

HIV infection

Coronary artery disease

Rheumatoid arthritis

Hematological malignancies (leukemias)

Besides preventing and treating inherited and infectious diseases, gene-transfer technologies probably will make possible the enhancement or replacement of genes that inßuence other traits such as height, weight, strength, stamina, and even intelligence. These capabilities will generate many questions about the regulation of such technologies and the fairness of access to these expensive protocols, as well as safety and privacy issues, among others.

"Pharming" animals to produce human drugs. Gene-transfer technologies already are being used to transfer human genes into farm animals such as sheep and goats for the purpose of generating large quantities of expensive human proteins for use as pharmaceuticals. (The process has been called "pharming.") The animals carrying human genes are called "transgenics" and are very difficult and expensive to develop. This situation has encouraged biotechnology companies to explore more efficient ways to reproduce the animals; cloning technologies such as those used to create the famous Scottish sheep Dolly and other cloned mammals like mice, goats, and cows are the results of these efforts. And a reasonable assumption is that many of the new reproductive technologies being perfected in our mammalian cousins will be effective in- and applied to-humans.

Xenotransplants: from pigs to people. Some 18,000 organ transplants take place each year, not nearly accommodating the 40,000 who wait for appropriate donors. Ten people die each day waiting for suitable human donor organs. Transplanting such organs as hearts and kidneys from genetically altered pigs and other animals into humans, a process called xenotransplantation, may have the potential to save lives. Current research is aimed at using DNA technologies to grow organs having human genes that make the organ's surface more "human like" and may help to minimize the chance for rejection upon transplantation into a human host. A concern is the unintended transfer of animal viruses to humans and the effects this might have beyond the patient to the population at large.

Identification

Multiple uses across species. DNA technology can be used to identify any type of organism, from humans and whales to plants, viruses, and bacteria. One important use is for identifying organisms contaminating soil, air, water, and food. Pinpointing a disease source in an epidemic, for example, is critical for its rapid control. These analyses are not limited to diseases affecting humans: they can be used to identify disease sources in livestock, poultry, and plants as well.

Some uses of human DNA identification are to establish paternity and other family ties in adoption and immigration cases, identify victims of wars and other catastrophes, and aid the courts in criminal cases where biological evidence (e.g., blood and sperm) is left behind. Interestingly, DNA data gathered from other species present at a crime scene, such as plants, dogs, cats, and viruses (HIV) also have been used as evidence in trials.

A controversial DNA databank. In July, police linked a dead Florida man's DNA to eight unsolved rapes in Washington, using only the data available from a national DNA databank, called CODIS (Combined DNA Index System). No other investigative leads were available. CODIS, which came online in late 1998, contains DNA descriptions, or "profiles," of offenders convicted of certain serious crimes. While many agree that this use of DNA technology can be of great benefit to society, one controversy surrounding DNA profiling stems from the potential of a DNA sample to reveal much more about an individual (and their family) than just their identity. While today's practices scan specific DNA regions that do not currently reveal such additional information, the human genome is still relatively unknown territory, and no one knows what types of information future technology may be able to uncover from stored samples.

Another source of concern over DNA databanking is the potential for expanding the use of databases beyond that originally intended. Thought-provoking historical examples of expanding database functions include the now pervasive social security number system that was originally started in the 1930s to help with a newly established retirement program, and the use of census records to round up Japanese-Americans for placement in interment camps during World War II.

AGRICULTURE AND ANIMALS

Stronger cotton, healthier livestock. For thousands of years people have modified traits in plants and animals indirectly through selective breeding. Today, our growing ability to directly alter an organism's genetic makeup, called genetic engineering, is having a major impact worldwide on agriculture and animal husbandry. A number of ongoing projects aim to decipher and manipulate the genomes of such economically important organisms as rice, corn, wheat, soy, cotton, sheep, goats, cows, pigs, and fish. Some of these explorations have led to the development of genetically modified plants that are providing higher yields, are more nutritious, and have increased resistance to herbicides, pests, and extremes of weather and temperature. In the United States this year, about half of all soybeans and a third of all corn planted were from genetically modified seeds, with most modifications aimed at pest and herbicide resistance.

Genetic alterations have produced ornamental crops such as carnations whose "aging genes" have been identified and turned off to allow an extended shelf life. Other plants are being genetically modified to produce biodegradable plastics, industrial oils and chemicals, low-calorie sweeteners, and human pharmaceuticals. Genetically modified animals are more nutritious and leaner, produce more milk, and are sometimes larger and more resistant to disease.

In a few recent examples, researchers reported adding rabbit genes to cotton plants to make the fiber as bright and soft as rabbit hair but stronger and warmer. A new strain of rice announced this spring contains a soybean gene for iron incorporation. This new rice can be used to treat the 30 percent of the world's population who are iron deficient and lack the means for expensive iron supplements.

Growing concerns. Consumer resistance to genetically modified plants and resulting foods, sometimes called "Frankenfoods," is strong in Europe and may be growing in the United States. Concerns center around environmental and consumer safety issues.

Particularly in the United Kingdom, the strength of resistance to genetically modified foods stems from a lack of trust in the government to protect its citizens, following the "mad cow" disease scare. Although genetically modified plants can decrease the use of pesticides and herbicides and thereby benefit the environment, a concern is that plants engineered to be more resistant to herbicides may pass on that trait through cross-pollination to related weed species in the wild. This could result in the creation of extremely resistant weeds requiring treatment with even more herbicides. Also, the impact of new pest-resistance traits on such nontarget organisms as visiting butterßies or birds is not known.

A potential health concern is that genes producing allergy-inducing proteins (such as those from peanuts) could be introduced into other food plants and consumers might unknowingly ingest a substance to which they could be allergic. (In the United States, the federal government is considering voluntary labeling of products derived from genetically modified organisms.) Another controversial issue is that genes introduced from one species into another may cause some consumers to violate religious restrictions against, for example, eating pork or beef.

A CAREFUL BALANCE

Genetic data and tools offer enormous potential benefits to humankind but pose significant risks as well. As the impact of the new genetics grows, we can expect the courts to be increasingly confronted with many novel, challenging, and sometimes disturbing issues. Scientific progress continues to advance rapidly as society scrambles to keep apace. But no one can anticipate some of the ways current and ever more powerful future DNA technologies will be put to use, nor their unintended and potentially controversial or adverse effects. As we begin to realize the benefits of the new genetics, maintaining a cautious approach will help minimize the risks.

- Article from Judicature.
- What Can the New Gene Tests Tell Us?-Article from the Judges' Journal of the American Bar Association.
- Bridging the Gap Between Life Insurer and Consumer in the Genetic Testing Era-Article from the Indiana Law Journal.

For what diseases are gene tests available?

Currently, more than 900 genetic tests are available from testing laboratories. Some gene tests available in the past few years from clinical genetics laboratories appear below. Test names and a description of the diseases or symptoms are in parentheses. Susceptibility tests, noted by an asterisk, provide only an estimated risk for developing the disorder. Contact GeneTests for comprehensive information on test availability and genetic testing facilities. Some Currently Available DNA-Based Gene Tests :

- Alpha-1-antitrypsin deficiency (AAT; emphysema and liver disease)
- Amyotrophic lateral sclerosis (ALS; Lou Gehrig's Disease; progressive motor function loss leading to paralysis and death)
- Alzheimer's disease* (APOE; late-onset variety of senile dementia)
- Ataxia telangiectasia (AT; progressive brain disorder resulting in loss of muscle control and cancers)

- Gaucher disease (GD; enlarged liver and spleen, bone degeneration)
- Inherited breast and ovarian cancer* (BRCA 1 and 2; early-onset tumors of breasts and ovaries)
- Hereditary nonpolyposis colon cancer* (CA; early-onset tumors of colon and sometimes other organs)
- Charcot-Marie-Tooth (CMT; loss of feeling in ends of limbs)
- Congenital adrenal hyperplasia (CAH; hormone deficiency; ambiguous genitalia and male pseudohermaphroditism)
- Cystic fibrosis (CF; disease of lung and pancreas resulting in thick mucous accumulations and chronic infections)
- Duchenne muscular dystrophy/Becker muscular dystrophy (DMD; severe to mild muscle wasting, deterioration, weakness)
- Dystonia (DYT; muscle rigidity, repetitive twisting movements)
- Fanconi anemia, group C (FA; anemia, leukemia, skeletal deformities)
- Factor V-Leiden (FVL; blood-clotting disorder)
- Fragile X syndrome (FRAX; leading cause of inherited mental retardation)
- Hemophilia A and B (HEMA and HEMB; bleeding disorders)
- Hereditary Hemochromatosis (HFE; excess iron storage disorder)
- Huntington's disease (HD; usually midlife onset; progressive, lethal, degenerative neurological disease)
- Myotonic dystrophy (MD; progressive muscle weakness; most common form of adult muscular dystrophy)
- Neurofibromatosis type 1 (NF1; multiple benign nervous system tumors that can be disfiguring; cancers)

- Phenylketonuria (PKU; progressive mental retardation due to missing enzyme; correctable by diet)
- Adult Polycystic Kidney Disease (APKD; kidney failure and liver disease)
- Prader Willi/Angelman syndromes (PW/A; decreased motor skills, cognitive impairment, early death)
- Sickle cell disease (SS; blood cell disorder; chronic pain and infections)
- Spinocerebellar ataxia, type 1 (SCA1; involuntary muscle movements, reflex disorders, explosive speech)
- Spinal muscular atrophy (SMA; severe, usually lethal progressive muscle-wasting disorder in children)
- Thalassemias (THAL; anemias-reduced red blood cell levels)
- Tay-Sachs Disease (TS; fatal neurological disease of early childhood; seizures, paralysis) [3/99]

Is Genetic Testing Regulated?

Currently in the United States, no regulations are in place for evaluating the accuracy and reliability of genetic testing. Most genetic tests developed by laboratories are categorized as services, which the Food and Drug Administration (FDA) does not regulate. Only a few states have established some regulatory guidelines.

This lack of government oversight is particularly troublesome in light of the fact that a handful of companies have started marketing test kits directly to the public. Some of these companies make dubious claims about how the kits not only test for disease but also serve as tools for customizing medicine, vitamins, and foods to each individual's genetic makeup. Another fear is that individuals who purchase such kits will not seek out genetic counseling to help them interpret results and make the best possible decisions regarding their personal welfare. More information on these questionable test kits is available from Dubious Genetic Testing, an online report provided by Quackwatch. For a brief overview of the current regulatory environment for genetic testing, see the Oversight of Genetic Testing, a Genetics Brief from the National Conference of State Legislatures.

Does Insurance Cover Genetic Testing?

In most cases, an individual will have to contact his or her insurance provider to see if genetic tests, which cost between $200 and $3000, are covered. Usually insurance companies do not cover genetic tests, those that do will have access to the results. Insured persons would need to decide whether they would want the insurance company to have this information. States have a patchwork of genetic-information nondiscrimination laws, none of them comprehensive. Existing state laws differ in coverage, protections afforded, and enforcement schemes. The National Conference of State Legislatures provides a listing of current legislation regarding genetic information and health insurance. The recent marketing of genetic test kits directly to consumers, may lead to an increase in demand for insurance coverage.

WHAT CAN THE NEW GENE TESTS TELL US?

A cartoon appearing almost half a century ago in The New Yorker featured a young boy watching his father review his report card. "What do you think the trouble with me is, Dad?" he asks with artful innocence. "Heredity or environment?" In one timeless scene, the cartoonist conveyed our fascination with genetics and the ongoing debate over just how much we can attribute to the genes we inherit from our parents. Lately we have learned a lot about our genetic legacy. We now know that, in fact, all diseases have a genetic component, whether inherited or resulting from the body's response to environmental stresses like viruses or toxins. The successes of the Human Genome Project (HGP) have even enabled researchers to pinpoint errors in genes--the smallest units of heredity--that cause or contribute to disease.

The ultimate goal is to use this information to develop new ways to treat, cure, or even prevent the thousands of diseases that afflict humankind. But the road from gene identification to effective treatments is long and fraught with challenges. In the meantime, biotechnology companies are racing ahead with commercialization by designing diagnostic tests to detect errant genes in people suspected of having particular diseases or at risk for developing them.

An increasing number of gene tests are becoming available commercially, although the scientific community continues to debate the best way to deliver them to a public and medical community that are unaware of their scientific and social implications. While some of these tests have greatly improved and even saved lives, scientists remain unsure of how to interpret many of them. Also, patients taking the tests face significant risks of jeopardizing their employment and/or insurance status. And because genetic information is shared, these risks can extend beyond them to their family members as well.

Even so, many more tests are in the works as dozens of new biotechnology companies vie to spin genetic data into gold. In the United States alone over four hundred laboratory programs aim to develop gene tests for disorders ranging from arthritis to obesity, and the list grows daily. The technology continues to advance rapidly, and future versions will allow simultaneous testing for hundreds of different genetic mistakes. The volume of available personal genetic data is on the brink of exploding, increasing the urgency of addressing ethical, legal, and social implications thereof. This was not unexpected. From its start over six years ago, HGP planners have dedicated at least 3 percent of the budget to grappling with just these issues.

Beginning with a short introduction to ground the reader in the DNA science underlying gene tests, this article explains some of the tests, their limitations, and the extraordinary potential of DNA medicine for the twenty-first century.

A GENETIC SCIENCE PRIMER

A gene is simply a piece of DNA, the chemical responsible for storing and transferring all hereditary information in a cell. Genes accomplish this by containing recipes for making proteins, the true workhorses of all our trillions of cells. All living organisms are made up largely of proteins, which provide the structural components of all our cells and tissues as well as specialized enzymes for all essential chemical reactions. Through these proteins, our genes determine how well we process foods, detoxify poisons, and respond to infections. Although our cells have the same genes, not all genes are active in all cells. Heart cells synthesize proteins required for that organ's structure and function, liver cells make liver proteins, and so on.

In humans and other higher organisms, a DNA molecule consists of two ribbon-like strands that wrap around each other, resembling a twisted ladder. The ladder rungs are made up of chemicals called bases, abbreviated A, T, C, and G. Each rung consists of a pair of bases, either A and T or C and G. We have three billion base pairs (six billion bases) of DNA in most of our cells; this is our genome. With the exception of identical twins, the sequence of the bases--the order of As, Ts, Cs, and Gs--is different for everyone, which is what makes each of us unique. Variation in base sequence, along with environmental factors, accounts for all our diversity, including disease. The DNA making up our genome is divided into tightly coiled packets called chromosomes, which reside in the nucleus of each cell. Each chromosome is a single DNA molecule, and lengths range from 50 million to 250 million bases. Scientists can distinguish the chromosomes by size, distinctive staining patterns, and other characteristics. Most cells have 46 chromosomes, 23 from each parent. A set of 23 contains 22 numbered chromosomes (1-22) plus either an X or Y sex-determining chromosome. Females receive an X from each parent, and males get one X and one Y. Sperm and egg cells only have 23 chromosomes, and mature red blood cells have none.

Chromosomes are not continuous strings of genes. Genes are interspersed among millions of bases of DNA that do not code for proteins (noncoding DNA) and whose functions are largely unknown. In fact, genes constitute only a tiny fraction of the human genome, a mere 3 percent. Scientists estimate that we have about 60,000 to 80,000 genes, whose sizes range from fewer than one thousand to several million bases. We have two copies of every gene, one from each of our parents.

FROM DIVERSITY TO DISEASE

For all our outward variation, we are surprisingly alike at the DNA level. Differences account for only one tenth of 1 percent of our DNA (about three million base pairs). Yet DNA base sequence variations are responsible for all our physical differences and influence many of our other characteristics as well. Sequence variation can occur in our genes, and the resulting different forms of the same gene are called alleles. People can have two identical or two different alleles for a particular gene. Variation also occurs outside the genes in the noncoding part of our DNA.

Mutations. While most DNA variation is normal, harmful sequence changes sometimes occur in our DNA that cause or contribute to disease. All DNA sequence changes--called mutations--are either passed down from parent to child (in the sperm or egg cells) or acquired during a person's lifetime. The vast majority of diseases are due to acquired changes, known as sporadic mutations. These mutations can arise spontaneously during normal functions, as when a cell divides, or in response to environmental stresses such as toxins, radiation, hormones, and perhaps even diet. Nature provides us with a system of finely tuned repair enzymes that find and fix most DNA errors. But as we age, our repair systems may become less efficient and allow us to accumulate uncorrected mutations. This can result in diseases such as cancer.

Depending on where in our genome they occur, mutations can have devastating effects or none at all. If they are small and fall in the vast sea of noncoding sequences, no one might be the wiser. Changes within genes, however, can result in faulty proteins that function at less-than-normal levels or those that are completely nonfunctional, causing disease.

Sometimes only a tiny change in DNA sequence will lead to a serious disease. The substitution of just a single base, for example, leads to sickle cell anemia. Other diseases are caused by deletions or additions of single or multiple bases. Too many repetitions of a particular sequence of three DNA bases can doom a person to Huntington's disease, a fatal neurological disorder; Fragile X syndrome, the most common form of inherited mental retardation; or myotonic dystrophy, a muscle-wasting disease. Other diseases can result from large rearrangements of DNA.

Single-Gene and More Complex Diseases. Some four thousand diseases are thought to be caused by a mutation in a single gene that is inherited from one or both parents. Most of these disorders are very rare, accounting for only about 3 percent of all disease. Some occur more frequently in particular ethnic groups. Among the more common inherited disorders for which single, causative genes have been identified are sickle cell anemia (African Americans and Hispanics), cystic fibrosis (Caucasians), and Tay Sachs (Ashkenazi Jews).

For most diseases the causes are much more complex. The common scourges afflicting Western civilization are thought to be due to a variety of gene mutations, perhaps acting together, or to a combination of genes and environmental factors. Heart disease, diabetes, hypertension, cancers, Alzheimer's disease (AD), schizophrenia, and manic depression are all examples of complex diseases.

Except for rare forms of these disorders that are inherited in some families, single mutated genes associated with complex diseases are not considered causative. Rather, they confer a susceptibility to their bearers and, given the right combinations of genes and environmental factors, will allow a disease to develop. Untangling the genetic and environmental contributions to complex disease will be one of the greatest challenges for medical researchers in the next century.

Finding Disease Genes. To find a gene that is a likely candidate for involvement in disease, scientists must search for DNA changes that are linked only with people who have a particular disease. Searching randomly through three billion base pairs of DNA for tiny changes that may be linked with disease is no easy task. Scientists labored through 10 years of tedious, painstaking work to find the genes for both Huntington's disease and cystic fibrosis. Thanks to the HGP, researchers now have some guidance from chromosome maps. Generated within the last two years, these maps specify thousands of unique DNA regions that act as mile markers along the chromosomal highways. These types of markers, which form a grid of known locations across every chromosome, are especially informative to researchers searching for small differences in DNA sequence among the members of large families. The high-quality maps have dramatically sped up the discovery of disease genes, reducing the hunt from years (at a cost of several million dollars) to months in some cases. Luck plays an important part in any gene hunt. Researchers studying large families with several cases of an inherited disease scan the genomes of all family members for any changes in marker DNA sequence that correlate with the presence of disease. How long it takes to find a disease gene this way depends in large part on the particular markers chosen.

In fall 1996, a region on chromosome 1 was found to be associated with a form of prostate cancer that runs in families. Researchers examined over 300 DNA marker regions in the genomes of some 100 families and compared the DNA sequences of affected individuals with healthy ones. The location of the implicated region containing the mutation was made available to the entire research community via the Internet. This region is now the focus of an intensive search for the causative gene by many groups around the world. Although the type of prostate cancer studied in these families is rare, researchers expect it will lead to insights into how the more common forms arise.

Once the disease genes themselves or their approximate chromosomal regions are finally identified, academic and commercial laboratories often translate these findings into gene tests that can detect the particular mutations associated with a disease.

Bibliography

Barnum, Susan R.: *Biotechnology: An Introduction*, Belmont, Thomson/Brooks/Cole, 2005.

Bauer MW: *Biotechnology-the Making of a Global Controversy*, Cambridge, Cambridge University Press, 2002.

Bernauer, Thomas: *Genes, Trade, and Regulation: The Seeds of Conflict in Food Biotechnology*, Princeton, Princeton University. 2003.

Broach, J.R.: *The Molecular Biology of the Yeast*, Cold Spring Harbor, Cold Spring Harbor Laboratory, 1981.

Chrispeels, Maarten : *Plants, Genes and Crop Biotechnology*, Sudbury MA, Jones and Barlett Publishers, 2003.

Clark, J.M.: *Experimental Biochemistry*, New York, W.H. Freeman and Company, 1977.

Conko, Gregory: *The Frankenfood Myth: How Protest and Politics Threaten the Biotech Revolution*, New York, Praeger, 2004.

David Sadava: *Plants, Genes and Crop Biotechnology*, Sudbury MA, Jones and Barlett Publishers, 2003.

Dixon, Dougal: *After Man-A Zoology of the Future*, New York, St. Martin's Press, 1981.

Dodds, John H.: *Plant Genetic Engineering*, New York, Cambridge University Press, 1985.

Drexler, K. Eric: *Engines of Creation: The Coming Era of Nanotechnology*, New York, Anchor Books, 1986.

Fransman M, Junne G, Roobeek A: *The Biotechnology Revolution?*, Oxford, Blackwell, 1995.

Friedberg, E.C.: *DNA Repair*, New York, WH Freeman and Company, 1985.

Fumento, Michael: *Bioevolution: How Biotechnology is Changing Our World*, San Francisco, Encounter Books, 2003.

George S. Paul: *Beyond Humanity: Cyber Evolution and Future Minds*, Roackland, Charles River Media, 1996.

Goodsell, David S.: *Bionanotechnology: Lessons From Nature*, Hoboken, Wiley-Liss, 2004.

King, R.B., Mihony, G., and Sheldon, J.K.: *Practical Environmental Bioremediation*, Lewis Publishers, Boca Raton, 1992.

Kurzweil, Ray: *The Age of Spiritual Machines*, New York, Penguin Books, 1999.

Moravec, Hans: *Mind Children: The Future of Robot and Human Intelligence*, Cambridge, Harvard University Press, 1988.

Morholt, E. : *Sourcebook for the Biological Sciences*, San Diego, Harcourt Brace JOvanovich, 1986.

Murray, David: *Seeds of Concern: The Genetic Manipulation of Plants*, Sydney, University of New South Wales, 2003.

Postman, Neil: *Technopoly: The Surrender of Culture to Technology*, New York, Vintage Books, 1992.

Primrose, S.B.: *Principles of Gene Manipulation*, London, Blackwell Scientific Publications, 1989.

Retzer, W.J.: *Biotechnology Workbook*, Englewood Cliffs, Prentice Hall, 1991.

Rifkin, Jeremy: *The Biotech Century*, New York, Penguin Putnam, 1998

Shetty, Kalidas: *Food Biotechnology*, New York, Dekker/CRC Press, 2005.

Smith, John E.: *Biotechnology*, Cambridge, Cambridge University Press, 2004.

Switzer, R.L.: *Experimental Biochemistry*, New York, W.H. Freeman and Company, 1977.

Towle, Albert: *Modern Biology*, Austin, Holt, Rinehart and Winston, 1988.

Index

H

I

M

O

P

R

S

T

□□□